Computational Modelling in Hydraulic and Coastal Engineering

Computational Modelling in Hydraulic and Coastal Engineering

Christopher G. Koutitas

Aristotle University, Thessaloniki, Greece

Panagiotis D. Scarlatos

Florida Atlantic University, Boca Raton, FL, USA

CRC Press
Taylor & Francis Group
Boca Raton London New York

CRC Press is an imprint of the
Taylor & Francis Group, an **informa** business

A SPON PRESS BOOK

CRC Press
Taylor & Francis Group
6000 Broken Sound Parkway NW, Suite 300
Boca Raton, FL 33487-2742

First issued in paperback 2019

© 2016 by Taylor & Francis Group, LLC
CRC Press is an imprint of Taylor & Francis Group, an Informa business

No claim to original U.S. Government works

ISBN-13: 978-0-4987-0891-3 (hbk)
ISBN-13: 978-0-367-87205-2 (pbk)

Library of Congress Cataloging-in-Publication Data

Koutitas, Christopher G., author.
  Computational modelling in hydraulic and coastal engineering / Christopher Koutitas and Panagiotis D. Scarlatos.
        pages cm
  Includes bibliographical references and index.
  ISBN 978-1-4987-0891-3 (hardcover : alk. paper) 1. Coastal engineering--Data processing. 2. Seashore--Mathematical models. I. Scarlatos, Panagiotis D., author. II. Title.

TC209.K69 2016
627.01'51--dc23   2015031038

Visit the Taylor & Francis Web site at
http://www.taylorandfrancis.com

and the CRC Press Web site at
http://www.crcpress.com

Contents

Authors

Christopher Koutitas graduated from the Department of Civil Engineering at Aristotle University of Thessaloniki (AUTh) in 1970. He attended the graduate program in water resources and earned an MSE from the Department of Civil and Geological Engineering at Princeton University in 1971, supported by a university scholarship. He earned a doctorate from the Department of Civil Engineering at AUTh in 1976.

Dr. Koutitas conducted research on coastal modelling from 1977 to 1979 as a post-doctorate research fellow on a British Council Fellowship in the Department of Civil Engineering at Manchester University.

He served as a professor of coastal and harbour engineering in the Department of Civil Engineering at Democritus University of Thrace from 1980 to 1986, as a professor of coastal and harbour engineering in the Division of Hydraulics and Environmental Engineering of the Department of Civil Engineering at AUTh from 1986 to 2014, and as a visiting professor at the Department of Civil Environmental and Geomatic Engineering at University College London during the academic year 2010 to 2011.

His main research activity area is computational modelling in coastal and harbour engineering, aiming at the technically and environmentally optimal design of coastal structures.

He has had more than 200 papers published in journals and international conferences, and he has written two relevant books, which have been translated into English.

He has taught undergraduate courses in coastal and harbour engineering and computational hydraulics and graduate courses in marine environmental protection and coastal zone management for more than 35 years. He organized and ran a European universities network within the Erasmus programme related to education on coastal engineering.

He participated as the scientific responsible in many research projects financed by national agencies and the European Commission (11th and 12th Directorate General). He contributed to the MAST program as an expert, participated in the Modeling Coordination Committee of the MAST program, and has had numerous consultantships with private enterprises as well as the public authorities of Greece in projects related to coastal zone management.

He served as member of the governing board of the National Centre of Marine Research of Greece (1985–1990); as chairman of the board of the Port Authority of Thessaloniki, Greece (2004–2006); and as vice chairman of the governing board of the International Hellenic University (2007–2010).

He is now a professor emeritus of AUTh.

 **Panagiotis D. Scarlatos** earned his diploma degree (1972) and doctorate degree (1981) in civil engineering at the Aristotle University of Thessaloniki, Greece. In 1989 he joined Florida Atlantic University (FAU) as a faculty member of the Ocean Engineering Department before moving to the Civil, Environmental and Geomatics Department, where from 2005 to 2013 he served as the department chairman. He is currently a professor in the same department and director of the Center for Intermodal Transportation Safety and Security.

Before joining FAU he worked as a research associate at Aristotle University of Thessaloniki, the University of Florida, and Louisiana State University. He also worked for the South Florida Water Management District as a staff water resources engineer.

He taught, conducted research and served on committees for a wide variety of water resources and environmental engineering areas. He has written more than 130 technical publications and received as principal investigator or co-principal investigator research funding of more than $3.3 million from local, state and federal agencies. He has served as an expert witness in a variety of national and international cases pertaining to water resources and related infrastructure.

He was awarded for two consecutive years a NATO-sponsored scholarship and was a Fulbright scholar research grantee.

Chapter 1

Introduction

1.1 MATHEMATICAL MODELS

Hydraulic and coastal engineers are faced with complex practical and theoretical problems that require extensive knowledge of hydrodynamics and related environmental issues such as pollutant spreading and sediment transport. In most cases, these problems can be adequately described and solved by means of mathematical models.

Mathematical models are tools widely used to quantify cause–response relations in a wide spectrum of applied disciplines, including engineering, biology, economics and social sciences. These models can be deterministic, stochastic or a combination of both, based on observational data and theoretical principles. The complexity and sophistication of the models can vary from a simple statistical equation to a complex system of nonlinear partial differential equations.

Mathematical models simulate the real world in a realistic but approximated manner. The approximate nature of all mathematical models is due to the simplifying assumptions and parameterizations necessarily made to reach a realistic mathematical formulation that accepts a feasible solution (Shiflet and Shiflet 2014).

For simple models, the solution can be exact, like an analytical closed-form solution, or approximate, like an open-form solution given as a series with infinite number of terms. For complicated models, the solution can only be obtained by numerical methods. Numerical analysis is the branch of mathematics that deals with the development and evaluation of numerical methods. Mathematical models that are solved numerically, mostly by engaging computers, are known as numerical models.

After the 1970s, extensive usage of numerical analysis and numerical modelling in hydraulic engineering led to the development of computational hydraulics (Vreugdenhil 1981; Brebbia and Ferrante 1983; Hromadka, Beech and Clements 1986; Abbott and Minns 1998). Computational hydraulics is part of the broader discipline known as computational fluid dynamics (CFD), which comprises of all branches of fluid mechanics but mainly industrial flows and geophysical flows. Nowadays, applications of computational hydraulics are ubiquitous. Proliferation of computational hydraulics was supported by the fact that major hydraulic laboratories in Europe and the United States switched from the cumbersome and expensive physical hydraulic models to research, development and standardization of mathematical-computational hydraulic models. Computational hydraulics can effectively and efficiently solve and analyse problems pertaining to water flow phenomena and design of hydraulic and coastal structures and address environmental implications of pollutant transport by advection, diffusion and dispersion (Chau 2010).

The numerical treatment of a mathematical model requires the synthesis of a solution algorithm. The word *algorithm* describes a predetermined sequence of basic arithmetic and logical operations for the solution of a mathematical problem, where from a set of input data X, a set of output (the solution in numerical form) data Y is produced.

$$Y = L(X) \tag{1.1}$$

where L is an operator. It is noticeable that most of the equations of hydraulic models belong to the category of linear homogeneous second-order partial differential equations (PDEs). Consequently the subject of computational hydraulics can provide a unified view and understanding of the various fields of hydraulic engineering, including flows in closed conduits, open channels and porous media (groundwater); waves and maritime hydraulics; and pollutant and sediment transport (Abbott 1991).

1.1.1 Framework of numerical modelling

Theories on mathematical models and their numerical solutions are quite extensive. An extremely synoptic presentation of their highlights is attempted in the following (Fletcher 1991).

1.1.1.1 Well-posed mathematical models

A mathematical model of a system is considered to be well-posed if it has the following:

- The solution algorithm produces a solution for all sets of input data under specified conditions and limitations.
- The produced solution is unique, that is, only one solution output corresponds to each set of input data.
- The output has to be related to the input, via a Lipschitz condition, that is, each infinitesimal change of input values (δX) results into a finite change of output (δY).

Also, for a mathematical model to be well-posed, it is necessary that the governing PDEs, the auxiliary data and the numerical algorithm are all well-posed.

1.1.1.2 Discretization and numerical solution of mathematical models

In numerical models the governing equations are reformulated in approximate manner, like when the differential equations are written as difference equations, by means of some numerical method. The solution domain is also appropriately discretized into one-, two- or three-dimensional cells, and the solution is approximated on the corner nodes, sides or the interior of the cells. Discretization of the continuous independent variables x, y, z and t into small steps Δx, Δy, Δz and Δt in combination with the approximation of the governing equations leads to truncation errors. Theoretically, the truncation error is eliminated when Δx, Δy, Δz and Δt tend to zero. However, by reducing the size of discretization steps, the number of computational steps increases, increasing the number or arithmetic operations to be performed, and subsequently the round-off error becomes significant. The balance between the truncation error and the round-off error usually leads to an optimal numerical solution that is approximating but not coinciding with the analytical solution (Figure 1.1).

Any numerical solution method needs to satisfy three conditions:

1. It has to be consistent, that is, the approximation used for the derivatives has to be correct, according to the numerical method used.

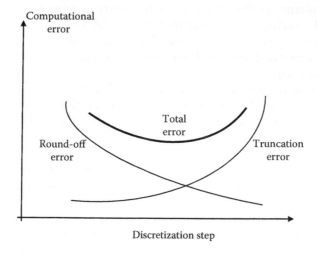

Figure 1.1 Numerical induced error versus the discretization step.

2. It has to be convergent, that is, the numerical solution must tend asymptotically towards the analytical solution, as the discretization steps (Δx, Δy, Δz, Δt) tend to zero. A non-convergent method is of no practical use.
3. It has to be numerically stable. For stable methods, the inevitably introduced errors during the solution procedure do not increase indefinitely but decay and become negligible after some solution steps.

1.1.1.3 Reliability of mathematical models

Using a numerical method and producing just a solution for a mathematical model is not the ultimate goal. A full application of a mathematical model requires involvement of three subsequent phases of (1) calibration, (2) verification and (3) validation. Model calibration is the quantitative determination of the model parameters. Model parameters are certain unknown variables and physical sub-processes that need to be identified a priori for the model to be operable. The variables are mostly expressed as lumped constants or known mathematical expressions. Regarding the sub-processes that are not described in detail by the model in order to avoid unnecessary computational complexity, they are approximated by parameters or relationships, taking specific values under specific conditions. Determination of the model parameters is based on available data sets of input–output values obtained from relevant physical models, field measurements or any available analytical solutions of the model. Those data are used only for the calibration phase of the model. A typical example of parameterization in fluid flows is the determination of the wall friction coefficient. This coefficient quantifies the effects of the boundary layer of the velocity profile. Thus, by excluding this layer from the model, the model solves for the bulk flow velocity outside of the boundary layer, while the boundary-layer effects are expressed by a wall-friction coefficient (e.g. Darcy-Weisbach friction coefficient).

Model verification is the proof of model truthfulness. Verification is conducted by using sets of known input–output data, different from those used for calibration, which should also be reproduced by the calibrated model.

Finally, model validation is the explicit recognition and delineation of the limits of model applicability, so that the users do not use the model outside those limits, because they may obtain non-realistic results. As already mentioned, the model formulation is based on simplifying approximations and parameterizations. Those assumptions impose (implicitly) the limits of model applicability, beyond which the assumptions for the model formulation are no longer valid. For example, if the model is based on the neglect of nonlinear terms (linearized model) the model is not valid for applications where the nonlinear terms become significant.

1.1.2 Development and application of numerical models

The correct development and application of numerical models is a lengthy process that requires competence and understanding of different scientific areas. Those areas as well as the requirements involved are described as follows:

- In-depth understanding of the physical phenomenon under consideration, through analysis and observations of field, laboratory and theoretical data.
- Mathematical representation of the physical phenomena involved by adopting a set of appropriate governing equations along with the necessary boundary and initial conditions.
- Making simplifying assumptions and conducting parameterization of the various variables involved.
- Selection of an appropriate numerical scheme for solution of the governing equations. Special attention should be given to the computational consistency, convergence and stability.
- Model calibration, verification and validation using the necessary data sets.
- Compilation, analysis and interpretation of the modelling results, followed by final reporting and presentation.

Lack of understanding or misinterpretation of a model's abilities and limitations can lead to inaccurate or even totally erroneous solutions.

1.2 FINITE DIFFERENCES METHOD

The method of finite differences is a classical method of numerical analysis, referring to the approximation of total or partial derivatives of functions, with respect to one or more free variables like x, y, z and t (the space and time variables), by divided differences of values of the functions.

The method is based on the expansion of a function in terms of Taylor series and is very useful for the numerical solution of PDEs and ordinary differential equations (ODEs) (Press et al. 1992; Cheney and Kincaid 2013). A function f(x) of independent variable, x, can be expanded in Taylor series in the vicinity of a value of x to obtain approximate values of the function at a nearby location $x + \Delta x$, or $x - \Delta x$ as

$$f(x \pm \Delta x) = f(x) \pm \frac{df}{dx}\Delta x + \frac{d^2f}{dx^2}\frac{(\Delta x)^2}{2!} \pm \frac{d^3f}{dx^3}\frac{(\Delta x)^3}{3!} + \dots \qquad (1.2)$$

where Δx is a small quantity. From Equation 1.2, the following equalities can be derived:

$$\frac{df}{dx} = \frac{f(x + \Delta x) - f(x)}{\Delta x} + O(\Delta x) \tag{1.3}$$

$$\frac{df}{dx} = \frac{f(x) - f(x - \Delta x)}{\Delta x} + O(\Delta x) \tag{1.4}$$

where $O(\Delta x)$ indicates that the truncation error is of the order Δx (due to the neglect of higher-order terms in the Taylor series). Thus the relations make possible the estimation of the first derivative of $f(x)$ at location x using values of $f(x)$ at a nearby location $x + \Delta x$ or $x - \Delta x$. Equation 1.3 is called the forward or upwind finite difference of the first order, and Equation 1.4 is called the backward or downwind finite difference of the first order.

By subtracting the two versions (upwind and downwind) of Equation 1.2 it can be easily induced that

$$\frac{df}{dx} = \frac{f(x + \Delta x) - f(x - \Delta x)}{2(\Delta x)} + O[(\Delta x)^2] \tag{1.5}$$

This relation is known as the central finite difference of the second order for the approximation of a first derivative. Notably its truncation error in Equation 1.5 is smaller than the error in the other two expressions (Equations 1.3 and 1.4) so it is more accurate.

By adding the two versions (upwind and downwind) of Equation 1.2 it yields

$$\frac{d^2 f}{dx^2} = \frac{f(x + \Delta x) - 2f(x) + f(x - \Delta x)}{(\Delta x)^2} + O[(\Delta x)^2] \tag{1.6}$$

This relation is also called the central finite difference of the second order for the approximation of a second derivative.

In the same way, starting with the expansion into Taylor series or similar approaches (as for example the approach of undetermined coefficients) we can approximate derivatives of a function by its values around a specific point (value of its free variable) in a consistent manner. The partial derivatives of multi-variable functions are also approximated in the same way, as they are simple derivatives of one variable, while the rest of the variables are kept constant (Faires and Burden 2015).

The approximation of derivatives by finite differences in an ODE or a PDE results in the local replacement of the differential equation by an algebraic equation valid for the specific location. That algebraic equation is formulated and is valid at a specific location of the solution domain of the equation and contains as unknowns the values of the function in a number of neighbouring points near $f(x)$, such as $f(x + \Delta x)$, $f(x - \Delta x)$, $f(x + 2\Delta x)$, $f(x - 2\Delta x)$ and so on. By discretizing the solution domain into a large number of fixed points, an equal number of algebraic equations can be derived. Solution of the system would provide numerical values of the function f on all those points. By definition, this is the numerical solution sought of the differential equation, that is, the determination of the arithmetic values of the unknown function $f(x)$ on a sufficient number of prefixed locations in the solution domain. This is in contrast to the analytic solution, which is defined as the estimation of the functional form of the solution $f(x)$ of the differential equation, so that its value at any x point can be subsequently calculated.

Depending on the procedure for the solution of the algebraic equations corresponding to all the points in the solution domain, the finite difference scheme used can be characterised as

- Explicit, when the deriving algebraic equations can be solved independently
- Implicit, when those equations need to be solved simultaneously as a system of algebraic equations

The locations where the numerical approximation of the differential equation is being sought are ordered regularly in the solution domain, usually at a constant in-between distance. This is done by discretizing the solution domain using one-, two- or three-dimensional orthogonal grids. The grid permits the decomposition of the solution domain into a number of cells and nodes easily characterised by coordinate indices, with positive integer values. For example, in a one-dimensional case at point $x_i = (i - 1)\Delta x$, from an origin at $x_1 = 0$, the function $f(x)$ is denoted as f_i, $f(x = x_i)$ or $f(x = (i - 1)\Delta x)$. Thus, if the solution domain for an ODE is $0 \le x \le 1$ and the domain is discretized into 10 segments ($\Delta x = 0.1$), then the unknown values of $f(x)$ would be $f_1(x = 0)$, $f_2(x = 0.1)$, ..., $f_{11}(x = 1.0)$. In order to compute those 11 values of the function $f(x)$, some of them need to be provided (i.e. boundary conditions: f_1 = known) and the rest need to be calculated by formulating and solving a number of algebraic equations equal to the number of the 10 remaining unknown values of $f(x_i)$.

The selection of the type of finite differences that will approximate the derivatives in a differential equation and the solution procedure of the group of the algebraic equations are not unique. In total they synthesise a solution algorithm or a numerical solution scheme.

From this discussion it can be concluded that the numerical solution is not always achievable, and when obtained, it may be useless. So, careful application of the methods and techniques of numerical analysis and post-solution checks are required to ensure that the millions of calculations (numerical solution) executed and the numbers produced by the computer have physical meaning and are operationally useful.

1.3 AIM AND STRUCTURE OF THIS BOOK

The material contained in this introductory book, *Computational Modelling in Hydraulic and Coastal Engineering*, is listed in the following chapters.

- Chapter 1 provides a brief introduction to mathematical modelling and sets the framework for the numerical treatment of differential equations with an emphasis on the method of finite differences.
- Chapter 2 describes the numerical treatment of ordinary differential equations through examples of reservoir storage management and water quality issues in a semi-enclosed lagoon.
- Chapter 3 provides an introduction to the classification of partial differential equations into elliptic, parabolic and hyperbolic, and presents simple examples of numerical schemes for solution of the Laplace equation (elliptic), the diffusion equation (parabolic) and the wave equation (hyperbolic).
- Chapter 4 presents the fundamentals of flow in pressurized conduits and delivers solutions for the classical Hardy Cross pipe network problem and the dynamics of water hammer.
- Chapter 5 deals with steady and unsteady free surface flows in one- and two-dimensional horizontal domains including geophysical stratified flows.
- Chapter 6 provides a brief description of surface gravity waves and delivers examples of propagation and modulation of those waves in one- and two-dimensional systems including the phenomenon of harbour basin resonance.
- Chapter 7 describes the mathematical description of groundwater flow in confined and unconfined aquifers, and provides numerical examples for one- and two-dimensional applications including saltwater intrusion.
- Chapter 8 deals with problems of mass transport of dissolved or particulate phase in one- or two-dimension domains, and delivers examples of transport of pollutants, sediment and air bubbles by using Eulerian and Lagrangian solution methodologies.
- Chapter 9 provides a brief description of other major numerical methods such as the weighted residuals, the finite elements method and the boundary integral method.

The use of the finite differences method is selected for the numerical solution of the mathematical models presented in the book for the following reasons:

- Its simplicity as a method of numerical analysis facilitates the educational scope of the book.
- The corresponding computer codes for its application are simple and easily conceived and assimilated by the reader, thus offering a direct opportunity for modification and extension by the reader.
- The processing power of current computers overshadows the benefits offered by the finite elements method, for example, in approximating irregular geometries of solution domains.
- Even today, operational computational fluid dynamics (CFD) codes used worldwide, especially for geophysical flows, are based on the finite differences method.

The book has been written for senior undergraduate students, graduate students and professionals working in the area of hydraulics and coastal engineering. Although a conscious effort has been made to include some introductory information on the different subjects covered, it is desirable that the reader has some background knowledge of and exposure to hydrodynamics, wave mechanics, differential equations, numerical analysis and computer programming using any of the commonly used languages (for example, MATLAB®, FORTRAN, C, C++ and BASIC).

There are 30 computer programs coded in MATLAB R2012b readily available for the reader to use. The programs are written with the novice rather than the expert in mind and special attention is given to maintain a simple code structure even at the expense of the full potential of MATLAB. Every computer application is followed by five practical problems for the reader to work and become more familiar with the application and the numerical model. It should also be noted that the code written for the graphics is suitable for the particular input data used; by changing the data, a slight modification of the code may be necessary for correct graphical output.

The reader needs to start by creating a MATLAB folder and downloading all of the files with extension .m or .mat into that folder. By opening MATLAB, all of those files should be visible on the left of the window within the Current Folder. Double-clicking on any of the MATLAB files (.m) will open the Script Editor on a new window and the program can be executed by clicking the Run arrow on top of the Script Editor. While the program is running, the Command Window in the middle of the MATLAB window will show progress. At the end of the simulation, the numerical data are available in the Workspace on the right section of the MATLAB window.

MATLAB was selected for the modelling for its widespread availability in academia and the industry, flexibility of use, extensive network user's support and excellent graphic capabilities (Chapman 2013; Moore and Sanadhya 2015).

Additional material is available at the CRC website: https://www.crcpress .com/product/isbn/9781498708913.

Chapter 2

Ordinary differential equations

2.1 WATER STORAGE RESERVOIR MANAGEMENT

The first application of the finite differences (FD) method is provided for the solution of an ordinary differential equation (ODE) that mathematically describes the filling and emptying of a water reservoir, such as an artificial lake behind a dam. The reservoir is filling from upstream catchment basins inflows, while at the same time being emptied by designed outflows over a weir, under a sluice gate or through an orifice.

Reservoir management involves several safety operational practices. Those practices include but are not limited to (1) maintaining a predetermined maximum water elevation for avoiding crest overtopping, (2) regulating water discharges for preventing downstream erosion or flooding, (3) estimating the outflow hydrograph for emergency evacuation and/or mitigation purposes, and (4) assessing the overall inflow–outflow storage responses of the reservoir to various hydrological inputs (Singh and Scarlatos 1988).

The storage capacity of a reservoir depends on the time-dependent water elevation, $z(t)$, and the corresponding horizontal water surface area, $S(z)$. The water storage, $V(z)$, of the reservoir can be quantified by the following integral:

$$V(z) = \int_0^z S(\zeta)\,d\zeta \tag{2.1}$$

In the limiting case of a cylindrical reservoir, $S(z) = S$ is a constant, and the water volume is calculated as $V(z) = S \cdot z(t)$. However for a dam blocking a river valley, the storage function $S(z)$ is usually described by a number of $S(z)$ or for discrete z values taken from topographic maps of the area. In that case, the volume of the reservoir is estimated numerically by using the trapezoidal rule of integration, that is, from the summation of the quantities $S(z) \cdot \Delta z$ for an $N + 1$ number of z values, from $z_0 = 0$ (bed) to the top

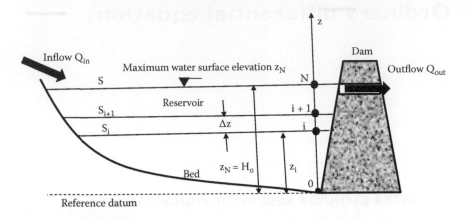

Figure 2.1 Schematic representation of reservoir operation.

water elevation $z_N = N \cdot \Delta z$ (Figure 2.1). The plot of $V(z_i)$ versus z_i is known as the rating curve.

$$V(z_N) = \frac{1}{2} \sum_{i=1}^{i=N} [S(z_{i-1}) + S(z_i)] \Delta z \tag{2.2}$$

The formulation of the mathematical model describing the volumetric changes of reservoir water as a result of inflow and outflow discharges can be derived by using the continuity equation and a stage–discharge relationship for the design outflow. The continuity equation implies that for an infinitesimal time (Δt) the difference between the inflowing ($Q_{in}\Delta t$) and the outflowing ($Q_{out}\Delta t$) volume of water will account for any accumulation or depletion of the reservoir water storage. Thus,

$$S(z)\Delta z = Q_{in}\Delta t - Q_{out}\Delta t \tag{2.3}$$

By assuming that outflow is occurring through an orifice of surface area A, then according to Torricelli's formula, the outflow depends on the hydraulic head (z) measured from the center of the orifice to the water surface,

$$Q_{out}^{orifice} = C_o A \sqrt{2gz} \tag{2.4}$$

where C_o is the discharge coefficient with an average value of 0.7.
  If outflow is occurring over a weir, then

$$Q_{out}^{weir} = C_w B(z - z_w)^{1.5} \text{ for } z > z_w, \text{ and } Q_{out}^{weir} = 0 \text{ for } z < z_w \tag{2.5}$$

where C_w is the discharge coefficient with an average value of 1.5, and z_w is the crest elevation of the weir measured from the same reference datum as the water elevation z.

Equation 2.3 leads to a difference-form relation, where the outflow is defined either by Equation 2.4 or Equation 2.5:

$$\frac{\Delta z}{\Delta t} = \frac{Q_{in}(t) - Q_{out}(t)}{S(z)} \tag{2.6}$$

For $\Delta z \to 0$ and $\Delta t \to 0$, Equation 2.6 can be written as an ordinary differential equation:

$$\frac{dz}{dt} = \frac{Q_{in}(t) - Q_{out}(t)}{S(z)} \tag{2.7}$$

The dependent variable z(t) is implicitly expressed in Equation 2.7 since both $S(z)$ and $Q_{out}(t)$ are functions of z. For solving Equation 2.6, it is required that the following data are known:

Geometric characteristics of the reservoir, $S(z)$
Inflow hydrograph, $Q_{in}(t)$
Geometric characteristics and discharge coefficient of the orifice (Equation 2.4) or weir (Equation 2.5)
Initial water elevation $z(t = 0) = H_o$

Achieving an analytical solution for Equation 2.7 is very unlikely, since most of the input data cannot be expressed in closed-form mathematical formulas. Thus Equation 2.6 is used instead, and the solution is accomplished numerically. For that purpose, the independent variable t is discretized by means of a time step Δt. Each discrete time value $t_n = n\Delta t$ is described by the subscript n, whereas the values of the other variables corresponding to that particular time are described by the superscript n, as $z^n = z(t_n)$ and $Q_{in}^n = Q_{in}(t_n)$.

2.1.1 Numerical solutions of the reservoir routing ordinary differential equation (ODE)

Using a forward finite differences scheme, and considering the case of reservoir emptying through an orifice, Equation 2.6 reads

$$\frac{z^{n+1} - z^n}{\Delta t} = \frac{Q_{in}^n}{S(z^n)} - \frac{C_o A \sqrt{2gz^n}}{S(z^n)} \tag{2.8}$$

which if solved for the unknown value z^{n+1} yields

$$z^{n+1} = z^n + \Delta t \left[\frac{Q_{in}^n}{S(z^n)} - \frac{C_o A \sqrt{2gz^n}}{S(z^n)} \right] \qquad (2.9)$$

The approximation of the ODE by Equation 2.9 for time level n + 1 is consistent and relates a new value of the water level z^{n+1} to a past value z^n, thus forming an explicit numerical solution scheme known as the Euler scheme.

An improved numerical approach is the Heun scheme where the advancement from time t_n to time t_{n+1} is calculated in two steps. If the quantities on the right-hand side are collectively abbreviated as F(z), then the method proceeds as follows:

$$z^{n+1} = z^n + \frac{\Delta t}{2} [F(z^n) + F(z^*)] \qquad (2.10)$$

where z^* is defined as $z^* = z^n + \Delta t \cdot F(z^n)$.

In addition to the Euler and Heun schemes there is a plethora of other numerical techniques including most notably the Runge-Kutta methods.

Since the initial value of $z(t = 0) = z^1$ is known, the numerical solution evolves with time successive computations of z^{n+1}. Simultaneously, the corresponding $S(z^{n+1})$ values are found by using the stage-surface (z-S) curve for the given reservoir. Since most of the time the stage-surface curve is discretized using a predetermined step Δz (Figure 2.2), the value of $S(z^{n+1})$ is estimated by linear interpolation as

$$\frac{S_{i+1} - S_i}{z_{i+1} - z_i} = \frac{S(z^n) - S_i}{z^{n+1} - z_i} \qquad (2.11)$$

Setting $\Delta z = z_{i+1} - z_i$ and $\lambda_1 = z^{n+1} - z_i$, the interpolated value of $S(z^n)$ from Equation 2.11 reads

$$S(z^n) = S_i + \frac{\lambda_1}{\Delta z} (S_{i+1} - S_i) \qquad (2.12)$$

At each time step, the numerical integration estimates a new z^{n+1} value, based on the previously computed value of z^n, as well as the known values of $S(z^n)$ and $Q_{in}(t^n)$. Thus, the solution produces a discrete time series of z(t) values, and, subsequently, the values of the outflow discharge Q_{out} and reservoir surface area S(z).

The computed values for all the time levels constitute the time series, readily applicable for operational use. Thus, the maximum achieved water

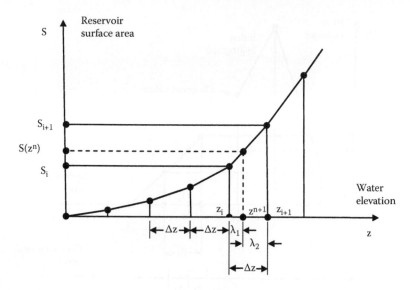

Figure 2.2 Interpolation from the stage–surface area curve.

stage in the reservoir during the filling–emptying process, the maximum value of the outflow discharge, the remaining water storage volume and other critical quantities can be easily estimated so that management decisions can be reached.

An interpolation process may be required for estimation of $Q_{in}(t_{n+1})$ if the inflow hydrograph is provided in discrete time steps δt, different from the computational time step Δt (Figure 2.3). Then, the value of $Q_{in}(t_{n+1})$ is estimated in terms of the measured values $Q_{in}(t_{m+1})$ and $Q_{in}(t_m)$:

$$Q_{in}(t_{n+1}) = Q_{in}(t_m) + \frac{\lambda_1}{\Delta t}[Q_{in}(t_{m+1}) - Q_{in}(t_m)] \tag{2.13}$$

where $\lambda_1 = t_{n+1} - t_m$.

Summarizing, the mathematical model consists of the ODE, the input parameters and the initial condition value for z. The solution algorithm consists of the following steps:

1. Provide input values for the
   a. Initial depth z^0
   b. Discharge coefficient C_o (or C_w)
   c. Inflow hydrograph $Q_{in}(t_m)$, for $t_m = m \cdot \delta t$
   d. Stage–reservoir surface relationship $S(z_i)$, for $z_i = n \cdot \Delta z$
   e. Computational time step Δt

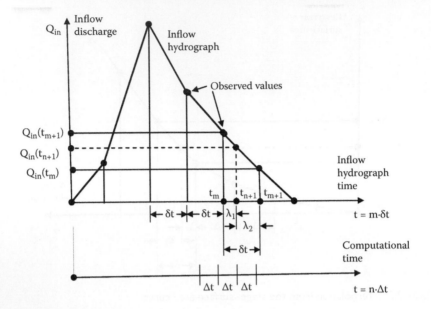

Figure 2.3 Interpolation of the inflow hydrograph.

2. Execution of the computations for the estimation of the series of z^n values (from $n = 1$ to $n = N_{max}$) and estimation of dependent parameters such as reservoir surface (S) and outflow discharge (Q_{out})
3. Storing of z, S and Q_{out} values in proper output data files
4. End of the algorithm

The following two practical problems involving the continuity equation (ODE) are solved numerically.

Example 2.1

A cylindrically shaped tank filled with water is emptying through a circular orifice at the bottom. Given the following data, the purpose of this exercise is to estimate the time required for the tank to empty:

Tank height = 10.0 m
Tank horizontal surface area = 5.0 m²
Orifice cross-section area = 0.005 m²
Discharge coefficient = 0.7

The numerical scheme is given by Equation 2.9 with no inflow $\left(Q_{in}^n = 0\right)$. For the numerical solution the tank is discretized vertically into N = 10 layers of equal thickness. Each layer (n) is confined between

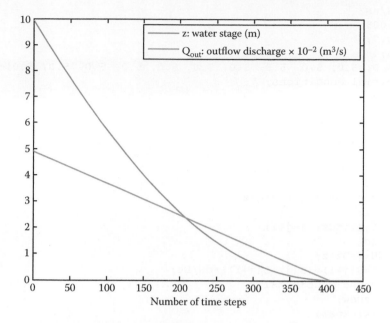

Figure 2.4 Water elevation of the emptying tank.

horizontal plane n and plane n + 1 (n = 1 to N + 1). The plane n = 1 corresponds to the bottom, and the plane n = N + 1 to the free surface. The computational time step is selected as 5 seconds. The simulation results of the receding water elevation and the outflow discharge are illustrated in Figure 2.4.

Computer code 2.1

```
% Example 2.1 Cylindrical Reservoir with No Refilling
Emptying through an Orifice
% S = Reservoir surface area of a cylindrical tank [m^2];
% H = Reservoir initial depth [m];
% A = Orifice surface area [m^2];
% Co = Discharge coefficient [dimensionless];
% Dt = Computational time step [s];
% Dz = Computational depth step [m];
% Qout(z) = Outflow discharge [m^3/s];
% z(t) = Numerical solution for the water depth [m];
% ns = number of horizontal sections;
clc; clear all; close all;
% Input data;
g=9.81;
ns=10;
H=10;
```

```
Dz=H/ns;
A=0.005;
Co=0.7;
Dt=5;
S=[5.0, 5.0, 5.0, 5.0, 5.0, 5.0, 5.0, 5.0, 5.0, 5.0, 5.0];
% Initial conditions;
z=H;
k=0;
% Main program;
for k=1:10000;
    if z<0
        z=0;
    end
    Qout=Co*A*sqrt(2*g*z);
    j=1;
    if z>j*Dz; j=j+1;
    end
    dH=j*Dz-z;
    S1=S(j+1)+(S(j)-S(j+1))*dH/Dt;
    znew=z-Qout*Dt/S1;
    z=znew;
    Z(k)=znew;
    % To facilitate plotting take 100*Qout;
    Qp(k)=Qout*100; m=k;
end
plot(1:m,Z,'b','Linewidth',1.5)
hold on
plot(1:m,Qp,'m','Linewidth',1.5)
v=[0, 450, 0, 10];
axis(v)
xlabel('Number of time steps'), ylabel('')
legend('z: Water stage [m]','Qout: Outflow discharge x 10E-2
[m^3/s]')
```

PROBLEM 2.1

Solve the same problem by making the following suggested changes while keeping the rest of the data constant:

1. Change the shape to a truncated cone-shaped tank of the same height but with a cross-section area of 1.0 m at the bottom and 11.0 m at the top, and plot the discharge Q_{out} as a function of time.
2. Change the shape to an inverted truncated cone-shaped tank of the same height but with a cross-section area of 11.0 m at the bottom and 1.0 m at the top, and plot the water elevation z as a function of time.
3. Change the orifice area from 0.005 m² to 0.01 m², and plot the temporal variation of both the discharge Q_{out} and the water elevation z.

4. Change the horizontal layers from 10 to 5.
5. Change the time step from 5 s to 50 s, and plot the discharge Q_{out} versus the water elevation z.

Compare the solution data obtained by running the modifications, derive conclusions and discuss the significance of the various variables involved to the estimation of the tank emptying time.

Example 2.2

A reservoir is receiving floodwaters from an upstream watershed. As a result, water storage increases, and when the water surface reaches a certain safety level, it discharges over a weir into the spillway. Given the following data, estimate the outflow hydrograph and the time variation of the water stage in the reservoir:

Initial reservoir water stage = 50.0 m
Weir crest elevation = 95.0 m
Weir width = 100.0 m
Varying reservoir horizontal area = 100 z^2; (z is the vertical height in metres)
Inflow hydrograph = $2000\left(1 - \dfrac{t}{T_d}\right)$; ($T_d$ = 40,000 s = 11.11 hr is the flood duration)

The numerical scheme is given by Equations 2.5 and 2.6. A computational time step of 100 s has been selected, and the calculations terminate when there is no discharge over the weir.

The simulation results of the inflow and outflow hydrographs as well as the water elevation are presented in Figure 2.5. The delay of the outflow discharge due to the filling of the reservoir is evident. Also, as it was expected, the peak of the water elevation was reached when the inflow and outflow rates were equal.

Computer code 2.2

```
% Example 2.2 Flood Routing Through a Reservoir
% Td = Flood duration [s];
% Qo = Initial flood discharge [m^3/s];
% zo = Initial water elevation in reservoir [m];
% zs = Spillway crest elevation [m];
% B = Spillway width [m];
% Dt = Time step [s];
% nm = Number of integration steps;
% ns = Values of reservoir surfaces at different elevations;
% nf = Values of flood hydrograph at different times;
clc;clear all;close all;
```

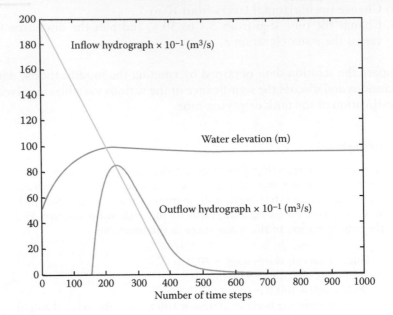

Figure 2.5 Water stage variation, and inflow and outflow hydrographs.

```
% Input data;
Td = 40000;
Qo = 2000;
zo = 50;
zs = 95;
B = 100;
Dt = 100;
nm = 1000;
z = zo;
t = 0;
% Initial conditions;
Qin = Qo;
Qout = 0;
% Main program;
for n = 1:nm;
    t = n*Dt;
    % Calculation of the inflow hydrograph;
    if Qin > 0
        Qin = Qo*(1-t/Td);
    else Qin = 0;
    end;
    % Calculation of the outflow discharge;
    if z > zs
        Qout = B*(z-zs)^1.5;
```

```
      else Qout = 0;
      end;
      % Reservoir horizontal plane area;
      SS = 100*z^2;
      % Calculation of the reservoir water elevation;
      z = z+Dt*(Qin-Qout)/SS;
      k = n;
      Qinp(k)= Qin;
      Qoutp(k)= Qout;
      zp(k)= z;
end;
plot(1:k,zp,'b','Linewidth',1.5);
hold on
plot(1:k,Qinp/10,'g','Linewidth',1.5)
plot(1:k,Qoutp/10,'m','Linewidth',1.5)
xlabel('Number of time steps')
text(100,180,'Inflow hydrograph x 10E-1 [m^3/s]')
text(370,52,'Outflow hydrograph x 10E-1 [m^3/s]')
text(500,105,'Water elevation [m]')
```

PROBLEM 2.2

Solve the same problem by making the following suggested changes while keeping the rest of the data constant:

1. Change the weir width from 100.0 m to 50.0 m, and plot the water stage variation in time.
2. Change the outflow from the weir (Equation 2.5) to the bottom outlet gate (Equation 2.4) and run the program until the closing of the outlet gate at the end of the inflow hydrograph $T_d = 40,000$ s (11.11 hr). Plot the inflow and outflow hydrographs.
3. Change the horizontal area function by assuming a reservoir of semi-spherical shape, and plot the water stage with time.
4. Change the inflow hydrograph using the following relations: 2500 $\left(\dfrac{t}{T_d}\right)$ for $0 < t < T_d$ and 5000 $\left(1-\dfrac{t}{2T_d}\right)$ for $T_d < t < 2T_d$. Plot the inflow and outflow hydrographs.
5. Change the time step from 100 s to 500 s and 1000 s, and plot the outflow hydrograph for the two new times steps.

Compare the solution data obtained by running these modifications, derive conclusions and discuss the significance of the various variables involved to the estimation of the outflow hydrograph and time variation of the water stage.

2.2 WATER QUALITY MANAGEMENT IN A LAGOON

A second application of the numerical solution of an ODE involves the water quality management in a lagoon connected through a small inlet to a vast water body (the open sea). The water volume of the lagoon, V, is renewed by means of a time-dependent discharge, Q(t). This inflow and (equal) outflow discharge is driven by various natural causes such as the wind, tide, watershed drainage and direct precipitation, and also human activities (i.e. pumping). Considering a time-average renewal discharge, $\bar{Q}$, then the renewal (or flushing) time of the lagoon is defined as $T_f = \dfrac{V}{\bar{Q}}$.

The water quality in the lagoon could be described by knowing the concentration levels of various substances, such as dissolved oxygen (DO), biochemical oxygen demand (BOD), nutrients, heavy metals and toxic organic compounds. In the following, for simplicity and without loss of generality, a single pollutant will be considered. Thus, the water quality will be determined by the mean concentration value (C) of a contaminant uniformly distributed over the entire lagoon area. Contaminants may enter the lagoon through point sources (industrial or municipal pipe outfalls) or distributed (agricultural or urban runoff). Considering a point source with flow rate q and contaminant concentration c, it is assumed that once discharged, the contaminant spreads and mixes instantaneously and uniformly through the lagoon.

In order to estimate the contaminant concentration, C(t), in the lagoon, a mass balance (continuity) equation can be derived for that particular substance. Using the volumetric approach, over a finite time interval (Δt), the volume of inflowing contaminant is $cq\Delta t$ and that of the outflowing is $C(Q + q)\Delta t$, assuming that there is not any pollution inflow from the open sea. The difference between the inflow and outflow contaminant fluxes defines the change of the contaminant concentration within the lagoon, $V\Delta C$. Thus, the continuity equation reads

$$V\Delta C = cq\Delta t - C(Q + q)\Delta t \qquad (2.14)$$

After taking the limits of ΔC and ΔQ, Equation 2.14 results in an ODE:

$$\frac{dC}{dt} = q\frac{c}{V} - (Q+q)\frac{C}{V} \qquad (2.15)$$

This is a first-order inhomogeneous ODE analytically solvable for constant q and Q, but requiring a numerical solution in the case of varying Q(t), or q(t) (or c(t)). If both q and Q are constant, after a considerable time the analytical solution for the concentration C stabilizes to

$$C = \frac{cq}{Q+q} \qquad (2.16)$$

In Equation 2.15, the last term clearly shows the significance of the flushing time, T_f. In case of a non-conservative (decaying) substance (e.g. microbial pollution) the contaminant is biodegrading at a rate $-\lambda C$, where λ is a decay coefficient with dimensions $(\text{time})^{-1}$ and the Equation 2.15 becomes (Scarlatos 2001)

$$\frac{dC}{dt} = q\frac{c}{V} - (Q+q)\frac{C}{V} - \lambda C \qquad (2.17)$$

The simplest numerical solution is the Euler scheme, where the derivative is approximated by a forward finite difference leading to the expression

$$\frac{C^{n+1} - C^n}{\Delta t} = q^n\frac{c}{V} - (Q^n + q^n)\frac{C^n}{V} - \lambda C^n \qquad (2.18)$$

Solving for the unknown value of concentration at time n + 1, the equation is written as

$$C^{n+1} = C^n + q^n\frac{c}{V}\Delta t - (Q^n + q^n)\frac{C^n}{V}\Delta t - \lambda C^n\Delta t \qquad (2.19)$$

This is an initial value problem, thus it requires knowledge of the contaminant concentration values at time t = 0.

Example 2.3

A lagoon is connected through a narrow inlet to a tidal sea. The discharge from an adjacent industrial pipe outlet contains a certain concentration of a decaying contaminant substance. The lagoon already contains some concentration levels of the same contaminant. In order to facilitate the cleaning process, in addition to the natural renewal (flushing) process, water is pumped out from the lagoon to the open sea (Figure 2.6). Assuming the renewal effects of the incoming and outflowing waters during the tidal cycle, the flushing flow is expressed as $Q = Q_0\left[\cos\left(\frac{2\pi t}{T}\right)\right]$ with $Q_{max} = Q_0$ and $Q_{min} = 0$. Given the following data, estimate the effects of three different pumping rates on the tidal-varied contaminant concentration within the lagoon:

Water volume of the lagoon = 50,000 m³
Initial contaminant concentration in the lagoon = 1.0 g/m³
Flow rate at the industrial site = 0.1 m³/s
Contaminant concentration of the industrial effluent = 500.0 g/m³
Tidal period = 43,200 s (semi-diurnal)

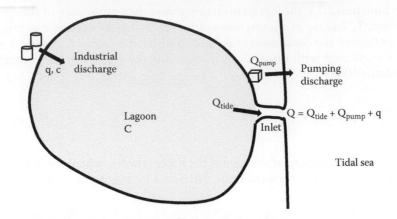

Figure 2.6 Water volume exchanges in the lagoon system.

Maximum tidal flow = 2.0 m³/s
Pumping rate = 0, 1.0 m³/s and 5.0 m³/s
Decay coefficient = 10^{-6} s⁻¹

The numerical scheme is given by Equation 2.19. A computational time step of 300 s has been selected and the calculations terminate after a pre-determined time period of 2000 time steps.

The tidal effects and the positive impact of pumping to the contaminant concentration as calculated by the numerical simulation are shown in Figure 2.7.

Computer code 2.3

```
% Example 2.3 Pollution in a Semi-enclosed Tidal Lagoon
% Qin = Pollutant discharge rate [m^3/s];
% Cin = Pollutant concentration rate at the source [g/m^3];
% Vol = Water volume of the lagoon [m^3];
% Qo = Maximum renewal discharge [m^3/s];
% Qout = Renewal discharge varying with the tide [m^3/s];
% Co = Initial pollutant concentration throughout the lagoon
[g/m^3];
% Rb = Biodegradation rate [1/s];
% Qp = Pump discharge rate [m^3/s];
% Tp = Tidal period [s];
% Dt = Time step [s];
clc; clear all; close all;
% Input data
Qin = 0.1;
Cin = 500;
Vol = 50000;
```

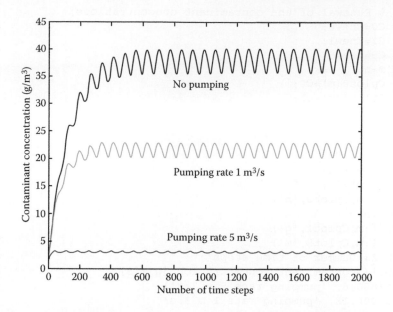

Figure 2.7 Pumping effects on contaminant concentration within the lagoon.

```
Qo = 2;
Co = 1;
Rb = 0.000001;
Qp0 = 0;
Qp1 = 1;
Qp5 = 5;
Tp = 43200;
Dt = 300;
C0 = Co;
C1 = Co;
C5 = Co;
% Main program;
for n = 1:2000;
    t= n*Dt;
    % Different outflow pumping rates;
    Qout0 = Qo*abs(cos((2*pi/Tp)*t))+Qp0;
    Qout1 = Qo*abs(cos((2*pi/Tp)*t))+Qp1;
    Qout5 = Qo*abs(cos((2*pi/Tp)*t))+Qp5;
    % Calculation of the contaminant concentrations;
    Cnew0 = C0 + (Cin*Dt*Qin/Vol) - (C0*Dt*Qout0/Vol)
- (Rb*Dt*C0);
    Cnew1 = C1 + (Cin*Dt*Qin/Vol) - (C1*Dt*Qout1/Vol)
- (Rb*Dt*C1);
    Cnew5 = C5 + (Cin*Dt*Qin/Vol) - (C5*Dt*Qout5/Vol)
- (Rb*Dt*C5);
```

```
% Renewal of the contaminant concentrations;
C0=Cnew0;
C1=Cnew1;
C5=Cnew5;
m=n;
Cplot0(m)=C0;
Cplot1(m)=C1;
Cplot5(m)=C5;
end
m=n;
Cplot0(m)=C0;
Cplot1(m)=C1;
Cplot5(m)=C5;
plot(1:m,Cplot0,'b')
hold on
plot(1:m,Cplot1,'g')
plot(1:m,Cplot5,'m')
xlabel('Number of time steps')
ylabel('Contaminant concentration [g/m^3]')
text(800,6, 'pumping rate 5 m^3/s')
text(800,18, 'pumping rate 1 m^3/s')
text(800,34, 'no pumping')
```

PROBLEM 2.3

Solve the same problem by making the following suggested changes while keeping the rest of the data constant:

1. Change the industrial discharge rate from 0.1 m^3/s to 5.0 m^3/s, and plot the contaminant concentration for six tidal cycles.
2. Change the decaying rate from 10^{-6} s^{-1} to 0.01 s^{-1}, and plot the difference of contaminant concentration between the two cases.
3. Change the initial concentration from 1.0 g/m^3 to 10.0 g/m^3, and estimate the pumping rate required to drop the maximum concentration level to less than 35 g/m^2.
4. Estimate the pumping rate required to reduce the maximum contaminant concentration in the lagoon below the value of 15.0 g/m^3 if the tidal effects are reduced by 50%.
5. For a zero pumping rate, estimate the necessary reduction in the contaminant concentration of the industrial effluent in order for the maximum concentration in the lagoon to be reduced by half.

Compare the solution data obtained by running these modifications, derive conclusions and discuss the significance of the various variables involved in the estimation of the contaminant concentration within the lagoon.

Chapter 3

Common partial differential equations of computational hydraulics

3.1 CLASSIFICATION OF PARTIAL DIFFERENTIAL EQUATIONS

Many phenomena in hydrodynamics are described by partial differential equations (PDEs) that cannot be solved analytically. As a result, solutions for those equations are feasible only by means of numerical algorithms. The general equation for linear PDEs of the second-order in two independent variables reads as follows:

$$A\frac{\partial^2 f}{\partial x^2} + B\frac{\partial^2 f}{\partial x \partial y} + C\frac{\partial^2 f}{\partial y^2} + D\frac{\partial f}{\partial x} + E\frac{\partial f}{\partial y} + Ff + G = 0 \tag{3.1}$$

where A to G are constant coefficients. Depending only on the coefficients of the second-order derivatives, a classification of those PDEs is accomplished based on the value of the discriminant:

$$\Delta = B^2 - 4AC \tag{3.2}$$

Thus, $\Delta < 0$ equations are classified as elliptic, $\Delta = 0$ as parabolic and $\Delta > 0$ as hyperbolic. This classification, in addition to being important from a mathematical point of view, has great significance for the hydrodynamic phenomena represented by the different classification groups. Under steady-state conditions (time-independent problems) the variables x and y are the spatial coordinates. Under unsteady conditions (time-dependent problems) the variable x is the spatial variable, while the variable y is replaced by the time variable t (Mitchell and Griffiths 1980; Lapidus and Pinder 1999).

Elliptic-type PDEs ($\Delta < 0$) in a two-dimensional solution domain are known as the Poisson equation, written as

$$\frac{\partial^2 f}{\partial x^2} + \frac{\partial^2 f}{\partial y^2} = \varphi(x, y) \tag{3.3}$$

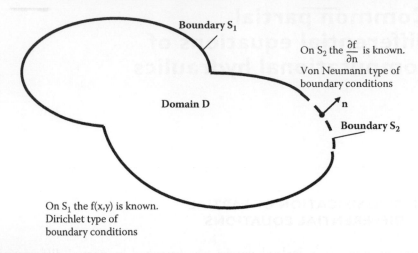

Boundary S_1

On S_2 the $\dfrac{\partial f}{\partial n}$ is known.

Von Neumann type of boundary conditions

Domain D

n

Boundary S_2

On S_1 the f(x,y) is known.
Dirichlet type of
boundary conditions

Figure 3.1 Solution domain and boundary conditions.

where φ is a known function. For φ = 0, the equation reduces to the well-known Laplace equation. In physical terms Equation 3.3 represents the distribution of a variable f(x,y) under steady-state conditions within a two-dimensional domain D, bounded by a curve S (Figure 3.1). For the problem to be well-posed, boundary conditions on the curve S must be known either in terms of the variable f(x,y) or the derivatives normal to the boundary $\dfrac{\partial f}{\partial n} = r$, where n is the direction normal to the boundary and r is a constant. Those boundary conditions are known as the Dirichlet and von Neumann, respectively.

For parabolic PDEs (Δ = 0), involving one-dimensional time-dependent problems, the equation in terms of the x-t variables takes the form

$$\frac{\partial f}{\partial t} = N \frac{\partial^2 f}{\partial x^2} \tag{3.4}$$

or for a two-dimensional space (variables x, y, t)

$$\frac{\partial f}{\partial t} = N\left(\frac{\partial^2 f}{\partial x^2} + \frac{\partial^2 f}{\partial y^2} \right) \tag{3.5}$$

where N is a constant. Equations 3.4 and 3.5 are known as the diffusion or heat equations. Physically they describe the diffusion process, that is, the spreading and simultaneous reduction in magnitude of a variable of known initial distribution $f(x, t = 0)$. The parameter N expresses the rate at which the diffusion is realized; faster for large and slower for small values of N. In order for the diffusion equation to be well-posed, both boundary conditions and initial conditions for the entire solution domain should be known.

For hyperbolic PDEs ($\Delta > 0$), describing one-dimensional time-dependent problems, the equation in terms of the x-t variables takes the form

$$\frac{\partial^2 f}{\partial t^2} = c_o^2 \frac{\partial^2 f}{\partial x^2} \tag{3.6}$$

or for a two-dimensional space (variables x, y, t)

$$\frac{\partial^2 f}{\partial t^2} = c_o^2 \left(\frac{\partial^2 f}{\partial x^2} + \frac{\partial^2 f}{\partial y^2} \right) \tag{3.7}$$

where c_o is known as celerity, phase velocity or speed of propagation. Equations 3.6 and 3.7 are known as wave or telegrapher's equations, and describe the propagation of a signal over time, in both positive and negative directions along the x-axis (and y-axis), moving from its original position ($f(x, t = 0)$) with celerity c_o. This equation applies to various types of water waves including gravity waves, tsunamis, astronomical tides, and elastic waves.

3.2 PARTIAL DIFFERENTIAL EQUATIONS AND CHARACTERISTIC DIRECTIONS

A much better appreciation of the physical significance of elliptic, parabolic and hyperbolic equations can be provided by the introduction of the concept of characteristic directions. Since the classification of PDEs depends only on the coefficients of the second-order derivatives, for simplicity Equation 3.1 can be condensed to

$$A \frac{\partial^2 f}{\partial x^2} + B \frac{\partial^2 f}{\partial x \partial y} + C \frac{\partial^2 f}{\partial y^2} + H = 0 \tag{3.8}$$

where H contains all of the remaining terms in Equation 3.1. If we consider within the solution domain a curve S on which the variable $f(x,y)$ and all of its derivatives satisfy Equation 3.8, then along a tangent to S the differentials $\dfrac{\partial f}{\partial x}$ and $\dfrac{\partial f}{\partial y}$ satisfy the relations

$$d\left(\frac{\partial f}{\partial x}\right) = \frac{\partial^2 f}{\partial x^2} dx + \frac{\partial^2 f}{\partial x \partial y} dy \tag{3.9}$$

$$d\left(\frac{\partial f}{\partial y}\right) = \frac{\partial^2 f}{\partial x \partial y} dx + \frac{\partial^2 f}{\partial y^2} dy \tag{3.10}$$

where $\dfrac{dy}{dx}$ defines the slope of the tangent to S. By combining Equations 3.8 to 3.10, the derivatives $\dfrac{\partial^2 f}{\partial x^2}$ and $\dfrac{\partial^2 f}{\partial y^2}$ can be eliminated leading to

$$\frac{\partial^2 f}{\partial x \partial y}\left[A\left(\frac{dy}{dx}\right)^2 - B\left(\frac{dy}{dx}\right) + C \right] - \left[A\frac{d}{dx}\left(\frac{df}{dx}\right) + H \right]\frac{dy}{dx} + C\frac{d}{dx}\left(\frac{df}{dy}\right) = 0 \tag{3.11}$$

By selecting a value of $\dfrac{dy}{dx}$ so that

$$A\left(\frac{dy}{dx}\right)^2 - B\left(\frac{dy}{dx}\right) + C = 0 \tag{3.12}$$

then Equation 3.11 reduces to

$$\left[A\frac{d}{dx}\left(\frac{df}{dx}\right) + H \right]\frac{dy}{dx} + C\frac{d}{dx}\left(\frac{df}{dy}\right) = 0 \tag{3.13}$$

The solutions of $\dfrac{dy}{dx}$ as derived from Equation 3.12 define the characteristic directions that apply to Equation 3.13. Thus, it can be easily seen from the discriminant Δ (Equation 3.2) that elliptic equations have no characteristic directions, parabolic have one, and hyperbolic have two. The characteristic directions are indicative of the way that information propagates within the solution domain, and it is very important in the numerical handling of the corresponding equations.

3.3 NUMERICAL SOLUTIONS OF TYPICAL PARTIAL DIFFERENTIAL EQUATIONS

In the following, numerical solutions of representative PDEs and associated hydraulic problems will be discussed. For simplicity, solutions are limited to applications involving one or two spatial dimensions (Rezzolla 2011).

The first step for the numerical solution is the discretization of the space–time solution domain by means of a regular grid with mesh sizes Δx, Δy and Δt. In the case of a two-dimensional space, the grids used, without any loss of generality, are square grids ($\Delta x = \Delta y$). As previously mentioned, the numerical solution is defined as the numerical calculation of the values of function $f(x,y,t)$ on some pre-determined points of the discretisation grid such as the nodes, the sides or the centres of the cells. This calculation is accomplished by discretizing the differential equation and solving the resulting algebraic equation (explicit scheme) or system of equations (implicit scheme) for the unknown values of the function on the grid points. Of course, due to the fact that boundary and initial conditions must be provided, the values of function $f(x,y,t)$ corresponding to those conditions would be known. For one-dimensional time-dependent flow, the boundary conditions (Dirichlet type) are given as known values of $f(x = 0, t)$ and $f(x = L, t)$ for all times t, where $x = 0$ and $x = L$ are the two ends of the spatial domain. The initial conditions are given as known values of $f(x, t = 0)$ for any point on the x-axis (Figure 3.2).

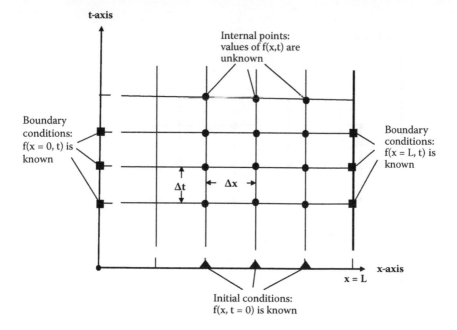

Figure 3.2 Discretization of the solution domain.

3.3.1 Solution of an elliptic partial differential equation (Laplace equation)

A two-dimensional elliptic PDE describes the distribution of a physical variable f(x,y) under equilibrium conditions. One simple expression of elliptic equations is the Laplace equation written as

$$\frac{\partial^2 f}{\partial x^2} + \frac{\partial^2 f}{\partial y^2} = 0 \tag{3.14}$$

The discretization of Equation 3.14 is accomplished by approximating the derivatives by the centred (by central) finite differences. The solution domain is discretized with a grid of mesh sizes Δx, Δy (commonly $\Delta x = \Delta y$), and the solution is calculated on the grid nodes, identified by the integer indices i,j, where $f_{i,j} = f(x_i,y_j) = f((i-1)\Delta x, (j-1)\Delta y)$ (Figure 3.3).

In the case of a common discretization step ($\Delta x = \Delta y$), the approximation of the derivatives by second-order central finite differences leads to the algebraic approximation of the differential equation

$$f_{i+1,j} + f_{i-1,j} + f_{i,j+1} + f_{i,j-1} - 4f_{i,j} = 0 \tag{3.15}$$

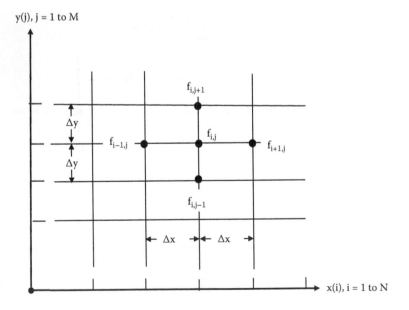

Figure 3.3 Identification of discretized function values.

or in terms of the unknown value $f_{i,j}$ as

$$f_{i,j} = \frac{1}{4}\left(f_{i+1,j} + f_{i-1,j} + f_{i,j+1} + f_{i,j-1}\right)$$

(3.16)

This simple algebraic relation suggests that the numerical solution scheme is implicit, requiring the synthesis and numerical solution of a system of algebraic equations, referring to all the (i,j) locations inside the solution domain. Equation 3.16 is not utilized at the boundary points where either the values $f(x,y)$ or the normal derivatives $\frac{df}{dn}$ are known. One special feature of the resulting system of algebraic equations is the sparseness of the constant coefficients matrix. The matrix is almost diagonal and can be easily solved by a successive iterations algorithm such as the easily programmable Gauss-Siedel iterative method.

3.3.1.1 Gauss-Siedel iterative method for solution of linear algebraic system of equations

If the upper index (k) annotates the order of the iteration, then the computation of $f_{i,j}^k$ values is accomplished starting from iteration (k = 1) and continuing with second, third, and so on iterations, until the following convergence criterion is satisfied:

$$\max_{i,j}\left|f_{i,j}^{k+1} - f_{i,j}^k\right| < \varepsilon \text{ for all } i,j$$

(3.17)

where ε is a very small predefined number. Using the notation in Equation 3.17, the Gauss-Seidel method can be described by the relation

$$f_{i,j}^{k+1} = \frac{1}{4}\left(f_{i,j+1}^{k+1} + f_{i-1,j}^k + f_{i+1,j}^{k+1} + f_{i,j-1}^k\right)$$

(3.18)

From Equation 3.18 it is evident that for the computation of $f_{i,j}^{k+1}$ the solution domain is swept from the smaller towards the bigger i,j values using the most recently corrected values of the function $f(x,y)$. This is an easily programmable procedure characterized by convergence to the correct value of f, not counting numerical errors, but with an unknown number of iterations. The convergence is secured when the diagonal coefficient is bigger or equal to the sum of the rest of the coefficients along a line of the coefficients matrix. In the case under consideration (Equation 3.15) the coefficient 4 is equal to the summation 1 + 1 + 1 + 1.

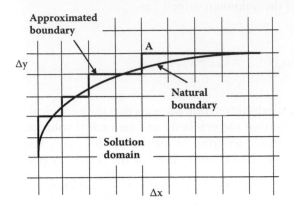

Figure 3.4 Approximation of a natural boundary.

3.3.1.2 Approximation of boundary conditions

The main challenge in the numerical solution of elliptic equations is the approximation of boundary conditions involving solution points on or near the boundaries. The computational speed of modern computers permits the discretization of the solution domain using a small discretization step, so the approximation of the boundaries of complex geometry does not present a problem (Figure 3.4).

As it was mentioned, through the boundary conditions the value of the function f(x,y) is given on the boundary, or its normal derivative $\dfrac{df}{dn}$ is specified. Under certain physical conditions, a situation arises where for part of the boundary values of f(x,y) are given, whereas for the rest of the boundary the normal derivative is specified. In the case of a given normal derivative $\dfrac{df}{dn}$ and according to the notations of Figure 3.5 the mathematical relation can be approximated as

$$\frac{\partial f}{\partial n} = \frac{f_A - f_B}{\Delta s} = \frac{f_A - f_B}{\Delta x \sqrt{1 + \lambda^2}} \tag{3.19}$$

Furthermore f_B can be expressed by linear interpolation in terms of f(x,y) values from the interior of the solution domain as

$$f_B = \lambda f_{i,j} + (1 - \lambda)f_{i+1,j} \tag{3.20}$$

Thus, the value of function f_A on the discretized boundary point A can be expressed by means of interior values of that function.

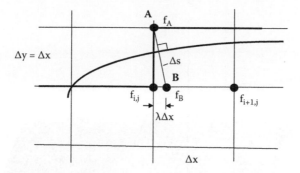

Figure 3.5 Approximation of von Neumann boundary conditions.

Example 3.1

A liquid enters a two-dimensional rectangular conduit and exits through an opening at the upstream (left side) end. The value of the velocity potential (f) is given at the upstream and downstream boundaries (Dirichlet type), while for the wall boundaries a no-flux condition is applied (Von Neumann type) (Figure 3.6). Considering a source (SS) located at the centre of the conduit, the purpose of this exercise is to calculate the distribution of the velocity potential throughout the solution domain:

   Potential at the upstream end = 10
   Potential at the downstream opening = 0
   Potential at the source = 10
   Upstream opening = 30 m
   Downstream opening = 5 m (opening is centred)
   Conduit length = 60 m

The physical phenomenon is governed by the Laplace equation (Equation 3.14). Thus the numerical algorithm used is that of

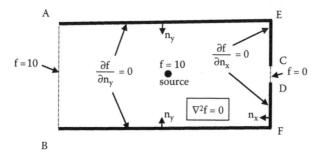

Figure 3.6 Solution domain and boundary conditions for conduit flow.

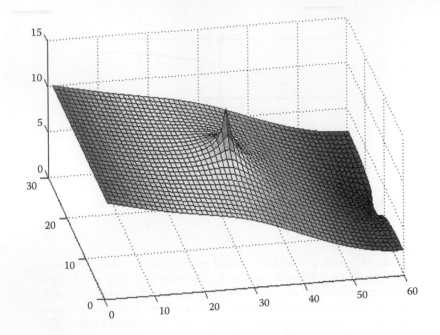

Figure 3.7 Distribution of the velocity potential.

Equation 3.18. The computational grid is defined as 30 × 60, and the number of iterations is set equal to 1000.

The distribution of the velocity potential along with the effects of the potential source and that of the boundary conditions are illustrated in Figure 3.7.

An interesting result of the simulation is how the computational error increases with increasing distance from the Dirichlet type of boundary conditions (Figure 3.8).

Computer code 3.1

```
% Example 3.1 Two-Dimensional Potential Flow in a
Rectangular Conduit
% f(i,j) = Velocity potential;
% SS = Potential source;
% im = Number of vertical nodes;
% jm = Number of horizontal nodes;
% iter = Number of iterations;
% fer = Difference in potential values between two
subsequent iterations;
clc; clear all; close all;
% Input data;
```

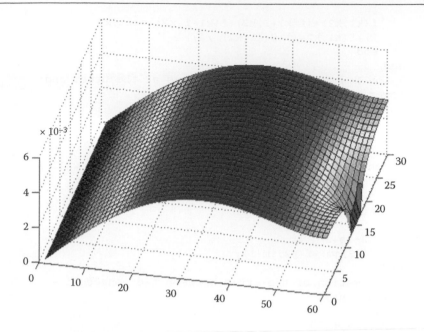

Figure 3.8 Distribution of the computational error.

```
im=30;
jm=60;
SS=10;
iter=1000;
% Initialization of the grid with zero potential values;
f(1:im,1:jm)=0;
is=im/2;
js=jm/2;
% Dirichlet type boundary conditions at the entrance and
exit of the conduit;
f(1:im,1)=10;
f(is-1:is+1,jm)=0;
k1=0;
k2=0;
% Spatial allocation of the potential source;
S(im,jm)=0;
S(15,30)=SS;
% Main program;
k=0;
for kk=1:iter
    k=1+kk;
    fold=f;
    % Estimation of the potential values;
    for k1=2:im-1
        for k2=2:jm-1
```

```
        f(k1,k2) = (f(k1+1,k2)+f(k1-1,k2)+f(k1,k2-
        1)+f(k1,k2+1)+S(k1,k2))/4;
        end
    end
    % Von Neumann boundary conditions at the upper and
    lower conduit walls;
    for j1=1:jm
        f(1,j1)=f(2,j1);
        f(im,j1)=f(im-1,j1);
    end
    % Von Neumann boundary conditions at the end wall of
    the conduit;
    for j2=1:(is-3)
        f(j2,jm)=f(j2,jm-1);
    end
    for j2=is+3 : im
        f(j2,jm)=f(j2,jm-1);
    end
    % Estimation of the error between two consecutive
    iterations;
    fnew=f;
    fer=fnew-fold;
end
i=1:k1+1;
j=1:k2+1;
fpp=fnew(i,j);
error=fer(i,j);
% Plot of the potential distribution;
surf(j,i,fpp);
figure
% Plot of the computational error distribution;
surf(j,i,error);
```

PROBLEM 3.1

Solve the same problem by making the following suggested changes while keeping the rest of the data constant:

1. Change the downstream boundary so that the potential is zero in sections EC and DF, while there is no flux in section CD (Figure 3.6), and plot the results.
2. Add a negative potential source of SN = −10 at the point with coordinates (15, 45), and plot the results.
3. For the sections EC and DF, define the Von Neumann relation as $\dfrac{\partial f}{\partial n_x} = 0.5$ and plot the results.
4. Knowing that the velocities are defined as $u_x = \dfrac{\partial f}{\partial x}$ and $u_y = \dfrac{\partial f}{\partial y}$, calculate the velocity field and plot the results.

5. Knowing that the streamline functions (s) are defined as $\dfrac{\partial s}{\partial y} = \dfrac{\partial f}{\partial x}$ and $\dfrac{\partial s}{\partial x} = -\dfrac{\partial f}{\partial y}$, calculate the streamline field and plot the results.

Compare the solution data obtained by running the modifications, derive conclusions and discuss the significance of the various variables involved to the phenomenon under consideration.

3.3.2 Solution of a parabolic partial differential equation (diffusion equation)

In contrast to the elliptic type PDEs that describe equilibrium conditions, the parabolic ones are time-dependent, therefore for a one-dimensional case the solution domain is the x-t plane. Considering a solution domain bounded by the inequalities $(0 \leq x \leq L; 0 \leq t \leq T)$, where L is the total length and T the total computational time, the discretization is accomplished by using a $\Delta x - \Delta t$ grid (Figure 3.8). Provided that the initial and boundary conditions are given, the unknown values of the f(x,t) can be calculated for the internal nodes of the computational grid. The values of f(x,t) on the computational points are denoted by the two integer indices. The lower one, i, refers to the position on the x-axis ($f_i = f(x_i)$, where $x_i = (i - 1)\Delta x$); and the upper index, n, refers to the position on the t-axis ($f^n = f(t^n)$, where $t^n = (n - 1)\Delta t$). Index i runs from 1 to N $((N - 1)\Delta x = L)$ and index n from 1 to M $((M - 1)\Delta t = T)$.

3.3.2.1 Forward in time, central in space (FTCS) numerical scheme

According to the aforementioned notations, the derivatives on point i,n are approximated by finite differences (FD) using a forward FD scheme of the first order for the time derivative $\dfrac{\partial f}{\partial t}$ (Equation 3.21) and a central FD scheme of the second order for the space derivative $\dfrac{\partial^2 f}{\partial x^2}$ (Equation 3.22):

$$\left[\frac{\partial f}{\partial t}\right]_i^n = \frac{f_i^{n+1} - f_i^n}{\Delta t} + \text{Truncation error } O(\Delta t) \tag{3.21}$$

$$\left[\frac{\partial^2 f}{\partial x^2}\right]_i^n = \frac{f_{i+1}^n - 2f_i^n + f_{i-1}^n}{(\Delta x)^2} + \text{Truncation error } O(\Delta x^2) \tag{3.22}$$

Then, after some simple rearrangement, the algebraic equation that approximates the differential equation (Equation 3.4) at point (i,n) reads

$$f_i^{n+1} = f_i^n + N\frac{\Delta t}{(\Delta x)^2}\left(f_{i+1}^n - 2f_i^n + f_{i-1}^n\right) \qquad (3.23)$$

This numerical algorithm is explicit and is known as the FTCS scheme. The numerical solution algorithm is organized as follows.

Note that in the above equation N is the diffusion coefficient, different from the upper limit N of the index n.

Using the known values of f(x,t) for n = 1 (for all i) and the known values of f(x,t) for i = 1 and i = N (for all n), the solution proceeds from time level n to time level n + 1 with the application of Equation 3.23 for i = 2 to i = N − 1 (Figure 3.9).

The FTCS scheme is consistent and convergent, but theoretical analysis and practical applications indicate that in order to ensure a stable numerical solution (without uncontrollable increase of numerical errors leading to a termination of the computations), the following inequality must be satisfied:

$$N\frac{\Delta t}{(\Delta x)^2} < \frac{1}{2} \qquad (3.24)$$

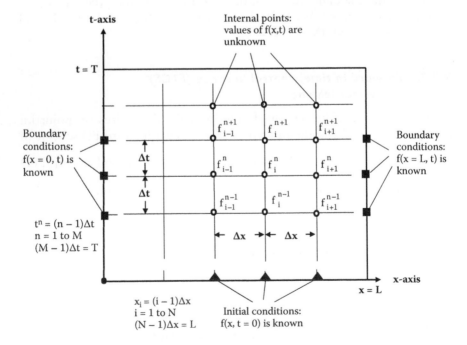

Figure 3.9 Identification of discretized variables.

Given the value of the diffusion coefficient N and the space discretization step Δx, the time step Δt has to be selected in compliance to the inequality Equation 3.24. Otherwise the numerical computations would overflow.

3.3.2.2 Crank-Nicolson numerical scheme

The limitation imposed using the FTCS scheme can be bypassed by employing an implicit scheme for the numerical solution of the parabolic equation. The FD used for the approximation of the time derivative is the same, but the space derivative is approximated by a second-order FD at time level n + 1/2 (i.e. the average of the FD at time level n and level n + 1)

$$\left[\frac{\partial^2 f}{\partial x^2}\right]_i^n = \frac{\left(f_{i+1}^n - 2f_i^n + f_{i-1}^n\right) + \left(f_{i+1}^{n+1} - 2f_i^{n+1} + f_{i-1}^{n+1}\right)}{2(\Delta x)^2} \tag{3.25}$$

The algebraic equation used for the computation of $f(x,t)$ values at n + 1 time level, after the isolation of all the f^{n+1} values on the left-hand side of the equation, has the form

$$-Rf_{i+1}^{n+1} + (1+2R)f_i^{n+1} - Rf_{i-1}^{n+1} = Rf_{i+1}^n + (1-2R)f_i^n + Rf_{i-1}^n \tag{3.26}$$

where $R = \dfrac{N\Delta t}{2(\Delta x)^2}$. On the right-hand side of the equation, all the $f(x,t)$ values are known from the previous time step, whereas on the left-hand side, the three consecutive values at the new time step are unknown.

Equation 3.26 cannot be used for a direct estimation of f_i^{n+1} as it was calculated in the explicit scheme (Equation 3.23), but all the algebraic equations written for the unknown f^{n+1} values have to be solved simultaneously as a system of equations. The solution applies only to the inner domain points, since the values of $f(x,t)$ on the boundaries are known.

The implicit numerical solution scheme leads to a tri-diagonal constant coefficients matrix, as three neighbouring f values are included in each equation. Its solution can be calculated either using an iterative method, like the Gauss-Seidel method, or, more efficiently, using the Thomas algorithm, which is the transformation of the tri-diagonal matrix to an upper diagonal one and back substitution. This numerical scheme (Equation 3.26), known as Crank-Nicolson, is unconditionally stable and does not impose any limiting relation between the Δx and Δt values selected for the discretization of space and time in the numerical solution.

Example 3.2

The phenomenon where a viscous fluid, confined between two flat plates, is set in motion by the relative movement of the plates is known as Couette flow. This flow is very important in engineering applications involving lubrication and friction between oscillatory machinery components. In the preceding, the velocity field and the shear stress at the oscillating plate are calculated. The solution is obtained under the assumption that the lower plate moves periodically with a velocity $u_{lower}(t) = U \sin \dfrac{2\pi t}{T}$, while the upper remains still. The data used for this example are as follows:

> Maximum velocity of the lower plate = 1 cm/s
> Period of oscillation = 500 s, 1000 s
> Kinematic viscosity of the fluid = 0.01 cm²/s, 0.1 cm²/s
> Distance between the plates = 19 cm

The governing equation of the physical phenomenon is the diffusion equation (Equation 3.4), where the function f represents the velocity distribution u(z, t). Thus the diffusion coefficient N is replaced by the kinematic viscosity (ν) and the spatial step Δx by Δz (vertical axis). The solution was accomplished by using the FTCS numerical scheme (Equation 3.23). The initial condition is set as u(z, t = 0) = 0; and the boundary conditions as u(z = 0, t) = 0 and u(z = H, t) = u_{up}(t), where H is the vertical distance between the plates. The expression used for estimating the shear stress is defined as $\tau_* = \dfrac{\tau}{\rho} = \nu \dfrac{du}{dz}$ (Newtonian fluid).

The effects of the oscillation plate period and the fluid viscosity on the shear stresses adjacent to the oscillating plate are shown in Figure 3.10. It is evident that the shear stress increases with increasing fluid viscosity.

Computer code 3.2

```
% Example 3.2 Two-Dimensional Couette Flow Caused by an
Oscillating Plate
% U = Velocity amplitude [cm/s];
% T = Oscillation period [s];
% v = Kinematic viscosity of the fluid [cm^2/s];
% Dt = Time step [s];
% Dz = Vertical step [cm];
% nm = Number of vertical discretization layers;
% tm = Number of time steps;
clc; clear all; close all;
% Input data;
U = 1;
T = 500;
T1 = 1000;
v = 0.01;
v1 = 0.1;
```

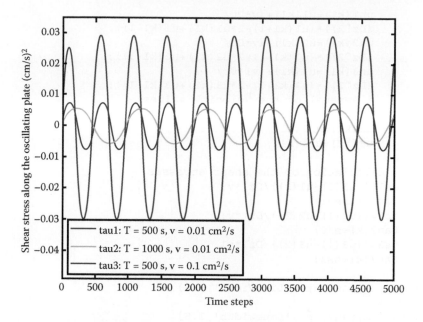

Figure 3.10 Fluid shear stress adjacent to the oscillating plate.

```
Dt = 1;
Dz = 1;
nm = 19;
im = nm +1;
tm = 5000;
% Initialization of the velocity field;
for m=1:im
    u1 (m) =0;  un1=u1 (m) ;
    u2 (m) =0;  un2=u2 (m) ;
    u3 (m) =0;  un3=u3 (m) ;
end
% Boundary condition at the fixed plate;
u1 (im) =0;
u2 (im) =0;
u3 (im) =0;
% Main program;
for k=1:tm
    Dtt=Dt*k;
    % Boundary condition at the oscillating plate;
    u1 (1) = U*sin (2*pi*Dtt/T) ;
    u2 (1) = U*sin (2*pi*Dtt/T1) ;
    u3 (1) = U*sin (2*pi*Dtt/T) ;
    % Calculation of the horizontal velocities between the
    two plates;
    for k1=2:nm
```

```
             un1 (k1) =u1 (k1) + (v*Dt/
             (Dz^2) ) * (u1 (k1+1) -2*u1 (k1) +u1 (k1-1) ) ;
             un2 (k1) =u2 (k1) + (v*Dt/
             (Dz^2) ) * (u2 (k1+1) -2*u2 (k1) +u2 (k1-1) ) ;
             un3 (k1) =u3 (k1) + (v1*Dt/
             (Dz^2) ) * (u3 (k1+1) -2*u3 (k1) +u3 (k1-1) ) ;
             % Renewal of the velocity values;
             u1 (k1) =un1 (k1) ;
             u2 (k1) =un2 (k1) ;
             u3 (k1) =un3 (k1) ;
         end
         % Calculation of the shear stresses;
         ss1= ( (u1 (1) -u1 (2) ) /Dz) *v;
         tau1 (k) =ss1;
         ss2= ( (u2 (1) -u2 (2) ) /Dz) *v;
         tau2 (k) =ss2;
         ss3= ( (u3 (1) -u3 (2) ) /Dz) *v1;
         tau3 (k) =ss3;
     end
end
plot (1:tm, tau1, 'r', 'Linewidth', 1.5)
hold on
plot (1:tm, tau2, 'g', 'Linewidth', 1.5)
plot (1:tm, tau3, 'b', 'Linewidth', 1.5)
xlabel ('Time steps'), ylabel ('Shear stress along the
oscillating plate [(cm/s)^2]')
axis ([0, 5000, -0.049, 0.035])
legend ('tau1:T=500s,v=0.01cm^2/s', 'tau2:T=1000s,v=0.01cm^2/
s', 'tau3:T=500s,v=0.1cm^2/s', 3)
```

PROBLEM 3.2

Solve the same problem by making the following suggested changes while keeping the rest of the data constant:

1. Calculate and plot the variation of the shear stresses halfway between the two plates.
2. Calculate and plot the velocity distribution for every 100 time steps starting at time of 3000 and ending at time 4000 time steps. Use v = 0.1 cm²/s and T = 500 s.
3. Assuming a non-Newtonian fluid $\tau_* = \varepsilon \left(\dfrac{du}{dz} \right)^2$, calculate the shear stress adjacent to the oscillating plate. Use $\varepsilon = 0.1$ and T = 500 s. Plot the results.
4. Calculate and plot the shear stress adjacent to the lower plate if the upper plate is oscillating in-phase with the lower plate but with half of its velocity. Use for the lower plate U = 1 cm/s and also v = 0.05 cm²/s and T = 500 s.

5. Calculate and plot the shear stress adjacent to the lower plate if the upper plate is oscillating out of phase (by 180 degrees) with the lower plate but with half of its velocity (U/2). Use for both plates T = 500 s and v = 0.05 cm²/s.

Compare the solution data obtained by running the modifications, derive conclusions and discuss the significance of the various variables involved in the phenomenon of Couette flow.

3.3.3 Solution of a hyperbolic partial differential equation (wave equation)

The linear hyperbolic equation of second order (Equation 3.6) physically describes the propagation of a wave or some other periodic signal. The signal is introduced at time t = 0, moving along both directions (positive and negative) of the x-axis (in the case of one dimensional space) with speed (wave celerity) equal to c_o.

The numerical solution is based on the finite differences approximation of the second derivatives (both spatial and temporal) by centred finite differences of second-order accuracy $O((\Delta x)^2, (\Delta t)^2)$. Using the same symbols regarding the f values on the discretized solution domain (Figure 3.9), the PDE is approximated on a point (i,n), that is, at a point with coordinates x_i, t^n, as follows:

$$\frac{\left(f_i^{n+1} - 2f_i^n + f_i^{n-1}\right)}{(\Delta t)^2} = c_o^2 \frac{\left(f_{i+1}^n - 2f_i^n + f_{i-1}^n\right)}{(\Delta x)^2} \tag{3.27}$$

The numerical scheme (Equation 3.27), known as the leap-frog scheme, is solved by using an algorithm similar to the one used for the case of the FTCS scheme for the parabolic equation. The algorithm starts from two initial time levels, with known values of the f(x,t) function (two-level initial conditions). The computation proceeds from one time level to the next, explicitly, as the terms in Equation 3.27 can be rearranged leaving on the left-hand side of the equation a single unknown term f_i^{n+1}:

$$f_i^{n+1} = 2f_i^n - f_i^{n-1} + c_o^2 \frac{(\Delta t)^2}{(\Delta x)^2}\left(f_{i+1}^n - 2f_i^n + f_{i-1}^n\right) \tag{3.28}$$

Equation 3.28 is readily applicable for time levels n > 2 on all the inner field points, as the end point values of $f(x_i, t^n)$, (for i = 1 and i = N) are the known boundary conditions (Figure 3.9). Since the numerical solution is explicit, to avoid numerical instability the value for the time step Δt is limited in accordance to the given values of Δx and c_o. Thus, the stability criterion that needs to be satisfied, known as the Courant-Friedrichs-Levi (CFL) criterion, reads

$$c_o \frac{\Delta t}{\Delta x} \leq 1 \tag{3.29}$$

A better insight of the numerical solution of the hyperbolic equations and the accompanying numerical errors can be accomplished by the numerical solution of the first-order hyperbolic equation known as the advective transport equation:

$$\frac{\partial f}{\partial t} + c_o \frac{\partial f}{\partial x} = 0 \tag{3.30}$$

Considering that

$$c_o = \frac{dx}{dt} \tag{3.31}$$

it is evident that the solution of Equation 3.31 is a family of straight lines $x = x_o + c_o t$, with slope c_o and x value at $t = 0$ equal to $x = x_o$. Combining Equations 3.30 and 3.31 yields

$$\frac{\partial f}{\partial t} + \frac{dx}{dt} \frac{\partial f}{\partial x} = \frac{Df}{Dt} = 0 \tag{3.32}$$

where D/Dt is the material or total derivative of f(x,t). That form implies that f(x,t) is constant along the characteristic curve $c_o = \dfrac{dx}{dt}$ (see Equation 3.12).

As shown in Figure 3.11, the solution consists of a plain translation of the initial form of f (initial condition f(x,t = 0)) along the positive x-axis direction with speed equal to c_o.

Thus, from the physical point of view, Equation 3.30 describes the propagation with speed c_o of a signal (wave) in the positive x-axis direction only, and also the dependence of the function f(x,t) on the values of the same function on preceeding space and time steps. Having the details of the analytical solution, different numerical solution schemes can be utilized and critically analysed.

3.3.3.1 Euler's unstable numerical scheme

When the time derivative is approximated with a forward FD and the space derivative with a central FD (FTCS), the algebraic approximation of the

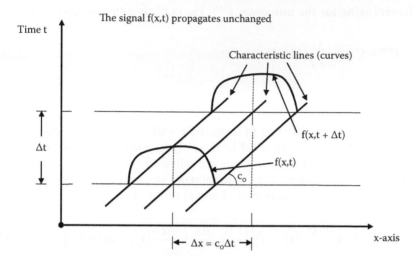

Figure 3.11 Characteristic lines in the x-t plane.

first-order hyperbolic equation at point (i,n) is known as the Euler's scheme in Equation 3.34 and has the form

$$\frac{f_i^{n+1} - f_i^n}{\Delta t} + c_o \frac{\left(f_{i+1}^n - f_{i-1}^n\right)}{2\Delta x} = 0 \tag{3.33}$$

Rearranging for the unknown f_i^{n+1}, the explicit solution reads

$$f_i^{n+1} = f_i^n - c_o \frac{\Delta t}{2\Delta x}\left(f_{i+1}^n - f_{i-1}^n\right) \tag{3.34}$$

The resulting numerical scheme (Equation 3.34) is consistent, but the solution, for any combination of Δx and Δt values, is unstable. The instability is the result of the fact that although the analytic solution depends only on the preceeding values in space and time of the function f(x,t), the above scheme in Equation 3.34 involves a preceding value in space that is f_{i+1}^n.

3.3.3.2 Godunov's numerical scheme

Using forward differences for the time derivative and backward differences for the space derivative (forward in time, backward space [FTBS]), the numerical scheme known as the Godunov's upwind scheme reads

$$\frac{f_i^{n+1} - f_i^n}{\Delta t} + c_o \frac{\left(f_i^n - f_{i-1}^n\right)}{\Delta x} = 0 \tag{3.35}$$

Rearranging for the unknown f_i^{n+1}, the explicit solution is written as

$$f_i^{n+1} = f_i^n - c_o \frac{\Delta t}{\Delta x}\left(f_i^n - f_{i-1}^n\right) \tag{3.36}$$

Equation 3.36 produces a numerical solution where the initial form of $f(x,t)$ is translated correctly with speed c_o, but the values of $f(x,t)$ undergo a certain degree of diffusion (decrease in magnitude with simultaneous spreading in space), as if they were controlled by a parabolic equation. Thus the solution presents a diffusive behaviour, as shown schematically in Figure 3.12.

However, the algebraic approximation of the equation can be re-written, after addition and subtraction of identical terms as

$$f_i^{n+1} = f_i^n - c_o \frac{\Delta t}{2\Delta x}\left(f_{i+1}^n - f_{i-1}^n\right) + \left(c_o \frac{\Delta x}{2}\right)\frac{\Delta t}{(\Delta x)^2}\left(f_{i+1}^n - 2f_i^n + f_{i-1}^n\right) \tag{3.37}$$

This is a finite differences algebraic approximation of a mixed-type equation, containing both the terms of the first-order hyperbolic and the second-order parabolic equations. Both the hyperbolic and parabolic parts are approximated by forward in time and central in space finite differences. The artificial diffusion coefficient of the parabolic part of the equation is equal to $c_o \frac{\Delta x}{2}$. Thus, when using forward time differences and backward space differences, instead of solving the hyperbolic equation, we actually solve a different equation and that fact introduces a numerical error known as numerical diffusion.

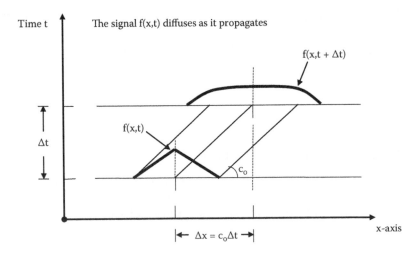

Figure 3.12 Numerical 'diffusion' of a propagating signal.

The implicit and unnoticed introduction, through the selected consistent FD scheme, of a diffusion term may suppress the anticipated numerical instability, but the effect is so dominant that it causes a diffusion of the f(x,t) values not compatible with the analytical solution of the hyperbolic equation. Notably, this numerical solution scheme, although inaccurate, is stable and viable. Also, it can be proven that the numerical scheme is subject to the CFL stability criterion (Equation 3.29). When the term $c_o \dfrac{\Delta t}{\Delta x}$ takes its limiting value (= 1), then the numerical solution is stable and coincides with the analytical solution (the numerical diffusion disappears).

The qualitative reasoning on the stability/instability of the numerical solution of the first-order hyperbolic equations presented in Equation 3.33 FTCS and Equation 3.36 FTBS can be documented mathematically by applying the von Neumann local stability analysis. According to this analysis, by assuming periodicity, the distribution of the computational errors ξ_j^n can be expressed as a finite complex Fourier series along grid points at the nth time step level. Thus, the initial error distribution reads

$$\xi_i^0 = \sum_{k=1}^{N-2} \alpha_k(t) e^{\sqrt{-1}(k\pi\Delta x)} \quad \text{where } i = 2,3,\ldots,N-1 \tag{3.38}$$

Then, the stability or instability is assessed by determining whether separate Fourier components of the error distribution will decay or amplify when the solution progresses to the next time level (n + 1).

For linear algebraic equations produced by finite difference discretization, the corresponding errors ξ_i^n satisfy the same homogeneous algebraic equations as the variable f_i^n. In addition, for those equations it is sufficient to study the propagation of the error of a single term $\xi_k^n = \alpha_k(t) e^{\sqrt{-1}(k\pi\Delta x)}$ of the Fourier series. Substitution of the error term ξ_k^n in Equation 3.33 leads to

$$\begin{aligned}
\alpha_k^{n+1} &= \alpha_k^n - \alpha_k^n \left(\frac{c_o \Delta t}{2\Delta x} \right) \left[e^{\sqrt{-1}(k\pi\Delta x)} - e^{-\sqrt{-1}(k\pi\Delta x)} \right] \\
&= \alpha_k^n \left[1 - \sqrt{-1} \left(\frac{c_o \Delta t}{\Delta x} \right) \sin(k\pi\Delta x) \right] = \alpha_k^n G_k^n
\end{aligned} \tag{3.39}$$

For the solution to be stable, the error needs to be contained. Thus the amplification factor G_k^n should be less than or equal to one. However, for Equation 3.33 the modulus of the amplification factor is given as

$$\left| G_k^n \right|^2 = 1 + \left[\left(\frac{c_o \Delta t}{\Delta x} \right) \sin(k\pi\Delta x) \right]^2 \geq 1 \tag{3.40}$$

and the numerical scheme is always unstable. By applying the same procedure to the Equation 3.36, the amplification factor is

$$G_k^n = 1 - \left(\frac{c_o \Delta t}{\Delta x}\right)[1 - \cos(k\pi \Delta x)] - \sqrt{-1}\left(\frac{c_o \Delta t}{\Delta x}\right)\sin(k\pi \Delta x) \qquad (3.41)$$

and its modulus reads

$$\left|G_k^n\right|^2 = 1 - 2\left(\frac{c_o \Delta t}{\Delta x}\right)\left[1 - \left(\frac{c_o \Delta t}{\Delta x}\right)\right][1 - \cos(k\pi \Delta x)] \qquad (3.42)$$

Therefore, the numerical scheme given by Equation 3.36 is stable when the RHS of Equation 3.42 is less than one, that is, when the CFL criterion (Equation 3.29) is satisfied.

3.3.3.3 Lax numerical scheme

To avoid instability induced by the second-order centred FD approximation of the space derivative, a number of numerical schemes have introduced an artificial controllable numerical diffusion. One of those methods is the Lax scheme, which is written in the following form:

$$\frac{f_i^{n+1} - f_i^n}{\Delta t} = -c_o \frac{\left(f_{i+1}^n - f_{i-1}^n\right)}{2\Delta x} + \frac{1}{2}\left(\frac{f_{i+1}^n - 2f_i^n + f_{i-1}^n}{\Delta t}\right) \qquad (3.43)$$

and leads to the explicit solution

$$f_i^{n+1} = \frac{1}{2}\left(f_{i+1}^n + f_{i-1}^n\right) - c_o \frac{\Delta t}{2\Delta x}\left(f_{i+1}^n - 2f_{i-1}^n\right) \qquad (3.44)$$

Equation 3.44 introduces an artificial diffusion term, which in combination with the centred FD for the space derivative produces a stable numerical solution as long as the CFL stability criterion is satisfied. Application of the Lax scheme results in propagation of a signal with the correct speed but with the formation of trailing oscillations that follow the main signal (Figure 3.13). That type of numerical error is known as numerical dispersion, and the name is derived from the fact that there is a differential rate of propagation of signals with different frequencies.

This can be verified using a composite signal that can be analysed to a number of Fourier (sinusoidal) components. Thus the dispersion would show through the differentiation of the speed of propagation of the various components with different frequencies. The highest frequency (small

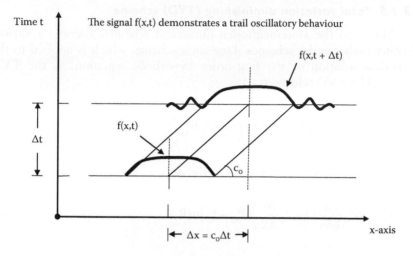

Figure 3.13 High-frequency oscillatory behaviour of the Lax numerical scheme.

wavelengths) components would propagate at lower speeds, forming a high frequency trail. Since the signal with the highest describable frequency is the wave with length $2\Delta x$, the trail has the form of spurious oscillations of that shortest wavelength following the main signal.

3.3.3.4 Fromm's numerical scheme

Another classical method is the Fromm scheme, which involves numerical dispersion but minimizes the numerical diffusion. This is an explicit scheme frequently utilized in the past for the solution of problems involving industrial fluid dynamics. Fromm's scheme is a five-point scheme written as

$$f_i^{*n+1} = f_i^n - c_o \frac{\Delta t}{4\Delta x}\left(f_{i+1}^n - f_{i-1}^n + f_i^n - f_{i-2}^n\right) \tag{3.45}$$

$$\begin{aligned}f_i^{n+1} = f_i^{*n+1} &- \left(c_o \frac{\Delta t}{2\Delta x}\right)^2\left(f_{i+1}^n - 2f_i^n + f_{i-1}^n\right) \\ &+ \left(c_o \frac{\Delta t}{2\Delta x}\right)\left[\left(c_o \frac{\Delta t}{2\Delta x}\right) - 1\right]\left(f_i^n - 2f_{i-1}^n + f_{i-2}^n\right)\end{aligned} \tag{3.46}$$

Other numerical schemes involve FD approximations of the space derivatives introducing numerical diffusion, controlled by a diffusion term with a negative diffusion coefficient. Such a term increases the function values.

3.3.3.5 Total variation diminishing (TVD) scheme

In addition to the aforementioned numerical schemes, there is a variety of other higher-order schemes. One such scheme, which is applied to the numerical solution of the first-order hyperbolic equation, is the TVD scheme. The TVD scheme reads

$$f_i^{n+1} = f_i^n - c_o \frac{\Delta t}{\Delta x}\left(f_i^n - f_{i-1}^n\right) - \left(K_{i+\frac{1}{2}} - K_{i-\frac{1}{2}}\right) \tag{3.47}$$

where the parameters involved are defined as

$$K_{i+\frac{1}{2}} = c_o \frac{\Delta t}{2\Delta x}\left(1 - c_o \frac{\Delta t}{\Delta x}\right)\left(f_{i+1}^n - f_i^n\right)\varphi(R_i) \tag{3.48}$$

$$K_{i-\frac{1}{2}} = c_o \frac{\Delta t}{2\Delta x}\left(1 - c_o \frac{\Delta t}{\Delta x}\right)\left(f_i^n - f_{i-1}^n\right)\varphi(R_i) \tag{3.49}$$

$$R_i = \frac{f_i^n - f_{i-1}^n}{f_{i+1}^n - f_i^n} \tag{3.50}$$

while $\varphi(R_i) = \min(2R_i, 2)$ for $R_i > 0$ and $\varphi(R_i) = 0$ for $R_i < 0$. It should be noted that if the parameters K are set equal to zero, then the TVD scheme coincides with the Godunov scheme.

Example 3.3

This exercise examines the performance of three different finite difference methods applied for solving the wave equation (Equation 3.30). The schemes applied are the Godunov scheme (Equation 3.36), the Fromm scheme (Equations 3.45 and 3.46) and the TVD scheme (Equation 3.47). The data used are as follows:

Celerity = 1 m/s
Time step = 0.5 s
Spatial step = 1 m

The upstream boundary condition for the function was set $f = 1$. The initial conditions were $f = 0$ for the Godunov and Fromm schemes and $f = 0.0001$ for the TVD scheme. The simulation was run for 300 spatial steps and 200 time steps.

From the numerical simulation results (Figure 3.14), the diffusiveness of the Godunov scheme, the trailing oscillations of the Fromm scheme and the robustness of the TVD scheme can all be clearly seen. In

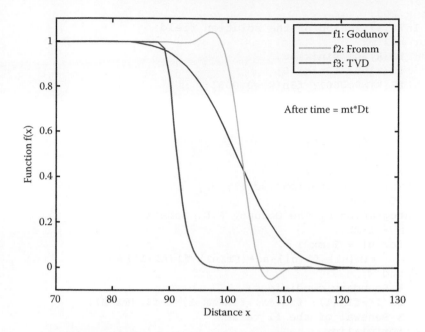

Figure 3.14 Simulation of pure advection by three finite difference schemes.

addition, since the simulation was performed for 100 s, the signal moving with a celerity of 1 m/s theoretically should have arrived at a distance of 100 m. To that effect, the TVD scheme was slightly delayed, the Fromm scheme was slightly ahead and the Godunov scheme, although highly diffused, travelled the right distance (Figure 3.14).

Computer code 3.3

```
% Example 3.3 One Dimensional Pure Advection
% C = Celerity [m/s];
% Dt = Time step [s];
% Dx = Spatial step [m];
% mt = Maximum time steps;
% mx = Number of steps along the x-axis;
% f1 = Solution using the Godunov scheme;
% f2 = Solution using the Fromm scheme;
% f3 = Solution using the TVD scheme;
clc; clear all, close all;
% Input data;
C = 1;
Dx = 1;
Dt = 0.5;
mx = 300;
```

```
mt = 200;
% Initialization of the solution field;
for k=2:mx
    f1(k)=0;  f1n(k)=0;
    f2(k)=0;  f2n(k)=0;
    f3(k)=0.0001;  f3n(k)=0.0001;
end
% Upstream boundary conditions;
f1(1)=1;  f1n(1)=1;
f2(1)=1;  f2n(1)=1;
f2(2)=1;  f2n(2)=1;
f3(1)=1;  f3n(1)=1;
% Stability criterion (sc<1);
sc = C*Dt/Dx;
% Integration by the Godunov F.D. scheme;
for n = 1:mt
    for n1 = 2:mx-1
        f1n(n1)=f1(n1)-sc*(f1(n1)-f1(n1-1));
    end
    % Boundary conditions;
    f1(1)=f1n(1);  f1n(mx)=f1n(mx-1)*2-f1n(mx-2);
    % Renewal of the f1 values
    for n1=1:mx
        f1(n1)=f1n(n1);
    end
    f1p=f1(1:n1);
    % Integration by the Fromm F.D. scheme;
    for n2=3:mx-1
        f2ni(n2)=f2(n2)-(sc/4)*(f2(n2+1)-f2(n2-1)+f2(n2)-
        f2(n2-2));
        f2n(n2)=f2ni(n2)+(sc/2)^2*(f2(n2+1)-2*f2(n2)+f2(n2-
        1))+((C*Dt)^2-2*C*Dt*Dx)/
        (4*Dx^2)*(f2(n2-2)-2*f2(n2-1)+f2(n2));
    end
    % Renewal of f2 values;
    for n2=1:mx
        f2(n2)=f2n(n2);
    end
    % Boundary conditions;
    f2n(mx)=f2n(mx-1)*2-f2n(mx-2);
    f2(1)=f2n(1);  f2(mx)=f2n(mx);
    f2p=f2(1:n2);
    % Integration by the TVD scheme
    % Estimation of the fr parameter;
    for n3=2:mx-1
        dd=f3(n3+1)-f3(n3);
        if dd == 0
            r1=-1;
        else
            r1=(f3(n3)-f3(n3-1))/dd;
```

```
                end
                if r1<0
                    fr=0;
                elseif r1>0 && r1<=1
                    fr=2*r1;
                else
                    fr=2;
                end
                akr=sc/2*(1-sc)*(f3(n3+1)-f3(n3))*fr;
                akl=sc/2*(1-sc)*(f3(n3)-f3(n3-1))*fr;
                f3n(n3)=f3(n3)-sc*(f3(n3)-f3(n3-1))-(akr-akl);
        end
        % Renewal of the f3 values;
        for n3=1:mx
                f3(n3)=f3n(n3);
        end
        % Boundary conditions;
        f3(1)=f3n(1);  f3(mx)=f3n(mx);
        f3p=f3(1:n3);
end
plot(1:n1,f1p,'r','Linewidth',1.5)
hold on
plot(1:n2,f2p,'g','Linewidth',1.5)
plot(1:n3,f3p,'b','Linewidth',1.5)
xlabel('Distance x'),ylabel('function f(x)')
axis([70,130,-0.1,1.1])
legend('f1:Godunov','f2:Fromm','f3:TVD')
text(110,0.7,'After time = mt*Dt')
```

PROBLEM 3.3

Solve the same problem by making the following suggested changes while keeping the rest of the data constant:

1. Change the celerity to 2.5 m/s and comment on the results.
2. Develop a computer code for the Euler finite difference scheme (Equation 3.34). Plot the data.
3. Change the boundary conditions from constant to sinusoidal, with amplitude equal to 1 and a period of 40 s.
4. Write the Godunov scheme in the form of Equation 3.37, and separately plot the hyperbolic and parabolic parts of the equation (second and third terms, respectively, in the right-hand side of the equation).
5. Develop a computer code for the Lax finite difference scheme (Equation 3.44). Plot the data.

Compare the solution data obtained by running the modifications, derive conclusions and discuss the significance of the various FD schemes involved in the phenomenon of pure advection.

Chapter 4

Flow in pressurized conduits

4.1 FUNDAMENTALS OF PIPE FLOW

In hydraulics, pipe flow is commonly considered as a full-pipe pressure-driven flow occurring in closed conduits of a circular cross-section. The two basic equations that describe pipe flow are the continuity (or mass conservation) equation and the energy equation. Assuming incompressible fluid and considering a pipe with varying cross-sectional area, the continuity equation can be written as

$$Q = u_1 A_1 = u_2 A_2 = \text{constant} \tag{4.1}$$

$$Q = u_1 \frac{\pi D_1^2}{4} = u_2 \frac{\pi D_2^2}{4} = \text{constant} \tag{4.2}$$

where Q is the volumetric discharge, u_i is the uniform cross-sectional velocity, and A_i and D_i are the cross-sectional area and pipe diameter, respectively, at point (i). However, in the majority of the applications the pipe cross-section is constant. Only in special cases (e.g. transitional sections connecting pipes of different diameter) does the pipe vary in diameter.

The energy equation is comprised of the kinetic part and the potential part (including the piezometric and the elevation energy). Written in terms of 'head', the energy equation reads

$$\frac{u_1^2}{2g} + \frac{p_1}{\gamma} + z_1 = \frac{u_2^2}{2g} + \frac{p_2}{\gamma} + z_2 + h_e \tag{4.3}$$

where p_i ($= \gamma y_i$) is the hydrostatic pressure at point (i), z_i is the elevation of the pipe from some reference datum, and h_e is the energy loss (head loss) between the two sections (Figure 4.1).

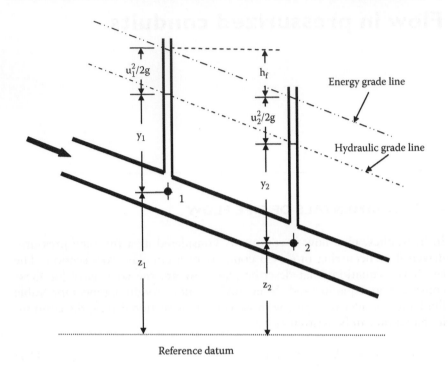

Figure 4.1 Pipe flow characteristics.

The term h_e includes both the frictional (h_f) and the local (minor; h_L) energy losses. The frictional losses are quantified by the well-known Darcy-Weisbach equation:

$$h_f = f \frac{L}{D} \frac{u^2}{2g} = f \frac{8}{g\pi^2} \frac{LQ^2}{D^5} = RQ^2 \tag{4.4}$$

where f is the friction coefficient and R is the resistance coefficient. The friction coefficient depends on the Reynolds number $R_e = \frac{\rho u D}{\mu} = \frac{uD}{\nu}$ and the relative roughness $\frac{\varepsilon}{D}$, where ρ is the fluid density, μ is the dynamic (or absolute viscosity), ν is the kinematic viscosity, and ε is the roughness of the pipe's interior surface (available from the manufacturer's specifications). The Darcy-Weisbach formula is valid for both turbulent and laminar flows. For laminar flow ($R_e < 4000$) the friction factor is independent of the relative roughness:

$$f = \frac{64}{R_e} \tag{4.5}$$

For intermediate turbulent flow ($4000 < R_e < 10^8$ and $0 < \dfrac{\varepsilon}{D} < 0.05$) the friction factor is quantified by the Colebrook equation as

$$\frac{1}{\sqrt{f}} = 1.14 - 2\log\left(\frac{\varepsilon}{D} + \frac{9.30}{R_e\sqrt{f}}\right) \tag{4.6}$$

For hydraulically smooth ($\varepsilon = 0$) turbulent pipe flow, the Colebrook equation reduces to the Prandtl equation:

$$\frac{1}{\sqrt{f}} = 2\log\left(R_e\sqrt{f}\right) - 0.8 \tag{4.7}$$

For high Reynolds numbers ($R_e > 10^8$) (hydraulically rough flow) the factor f depends only on the relative roughness and is expressed by the von Karman equation as

$$f = \left[1.14 - 0.869\ln\left(\frac{\varepsilon}{D}\right)\right]^{-2} \tag{4.8}$$

It should be noted that the Colebrook and the Prandtl equations are implicit and require an iteration process for the estimation of the friction factor (f).

Minor losses occur locally for reasons such as pipe expansion or contraction, pipe bends and valves. In general those losses are expressed as

$$h_L = K\frac{u^2}{2g} \tag{4.9}$$

where K is some experimental parameter.

4.2 PIPE NETWORK

Although most pipe flow problems are easy to solve, the one that imposes a computational challenge is the municipal water distribution pipe network. Pipe networks are comprised of a large number of interconnected pipes forming loops and branches. The pipes (branches) are connected on the junctions (nodes) of the network, and in groups they form closed loops. For simplicity it is assumed that pipe flow is steady and that water delivery to consumers occurs only from the junctions. Given the discharges inflowing or outflowing from the network junctions, the unknown quantities are the discharges of all pipes connected at the junctions. For this purpose the continuity and energy equations are employed (Lencastre 1995).

4.2.1 Continuity (junction) equation

Due to the incompressibility of water, at every junction (j) the sum of all discharges – inflowing or outflowing and those from the connected pipes (k-number) – should be zero:

$$\sum_k (\sigma_k Q_k) \pm Q_j = 0 \tag{4.10}$$

where Q_k are the flows of all connecting pipes at joint j, $\sigma_k = (\pm)$ accounts for the sign of each pipe flow (positive for inflowing and negative for outflowing) and Q_j is the inflowing (+) or outflowing (–) discharge at the junction (j = 1 to J-number of junctions; also see Figure 4.2). Since both the direction and magnitude of the discharge of the connecting pipes are unknowns, it is important that the discharges selected during the first approximation satisfy the continuity equation (Equation 4.10), and also the convention of positive and negative flows should be applied consistently throughout the network.

4.2.2 Energy (loop) equation

As it has been presented before, for each branch of the network (i), the energy loss h_{f_i} depends on the resistance coefficient R_i and Q_i^2 (where i = 1,

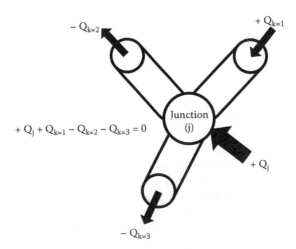

Figure 4.2 Continuity of flows at a junction.

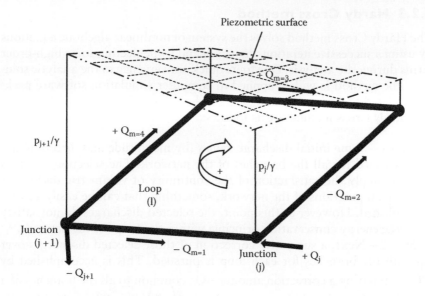

Figure 4.3 Energy balance within a loop.

N-number of branches; Equation 4.4). Thus, an equation based on the energy conservation principle can be written for each loop of the network, stating that the algebraic sum of the head losses along the branches of a certain loop is zero. This is true when no pump or turbine is placed along a pipe that would add or subtract energy from the loop. The loop equation can also be explained by the fact that the total head loss between any two junctions is independent of the path followed (Figure 4.3). Thus, for each loop (l) the energy equation takes the form

$$\sum_m \left(\sigma_m^* R_m Q_m^2 \right) = 0 \tag{4.11}$$

where σ_m^* is the directional sign of the flow in each branch forming the loop (m is the number of branches comprising loop l). In the following, the clockwise direction is taken as positive and the convention is retained for all the loops of the network.

The total number of unknown discharges equals the number of N-branches in the network. The total number of equations that can be written is (J − 1)-number of junction equations (Equation 4.10) and M-number of loop equations (Equation 4.11). Since N = M + J − 1, the number of unknowns equals the number of equations. The resulting non-linear system can also be solved numerically by using some successive iteration algorithm.

4.2.3 Hardy Cross method

The Hardy Cross method solves the system of nonlinear algebraic equations by using a successive iteration algorithm and by neglecting some high-order terms during the analysis. Due to its quick convergence to the analytic solution, the method is widely applied in computer simulation software packages related to hydraulics and structural analysis. The sequential steps of the Hardy Cross method are as follows:

   Step 1—Some initial discharge values (by magnitude and direction) are selected for all the branches of the network. The selection criterion is simply the satisfaction of the continuity of volumetric discharges on all junctions of the network, something that can be easily accomplished. However at this point, the selected discharges do not satisfy the energy conservation principle within the loops of the network.
   Step 2—Next, a successive correction of those selected discharges over all the branches for each loop is pursued. This is accomplished by estimating a correction amount ΔQ_l, common to all the branches of a loop (l), assigned according to the sign $\left(\sigma_m^*\right)$ of each branch (relatively to the loop it belongs). If the number of iterations is (n) then the next iteration (n + 1) produces a corrected discharge of the branch (m) according to the relation

$$Q_m^{n+1} = Q_m^n + \sigma_m^{*n}\Delta Q_l \tag{4.12}$$

   The correction ΔQ_l is found using the energy conservation equation along each loop as

$$\sum_l \sigma_m^* R_m \left(Q_m + \sigma_m^*\Delta Q_l\right)^2 = 0 \tag{4.13}$$

where m is the number of branches that form the loop. Assuming that ΔQ_l is small as compared to Q_m and neglecting the second-order term ΔQ_l^2, then Equation 4.13 yields

$$Q_l = \frac{\displaystyle\sum_l \sigma_m^* R_m Q_m^2}{\displaystyle\sum_l 2R_m Q_m} \tag{4.14}$$

since $\sigma_m^*\sigma_m^* = 1$. This correction, given the proper sign, is added algebraically to Q_m of the branch estimated during the previous iteration step. Of course, after the correction along one loop, the corrected discharges on all neighbouring loops (having common branches with the recently 'corrected' one) no longer satisfy the energy balance. Thus a new correction must be made for all the loops.

Step 3—The procedure is repeated over all loops as many times as needed until it satisfies a convergence criterion, that is, the maximum value of correction ΔQ_l for all loops of the network is less than a predefined small number.

During the solution process it should be emphasized that (1) after each correction along a loop the volumetric continuity principle is maintained on all junctions of the network, and (2) in the case that a discharge of a certain branch changes sign due to the imposed correction, then care should be taken for changing the sign of the discharge in the subsequent corrections. This iterative procedure can be conveniently coded for computer programming, provided that appropriate information is given describing the structure of the network. This is accomplished by the synthesis of the connectivity matrix having (M) rows and (N) columns equal to the M-number of loops and the N-number of branches, respectively. Each element of the matrix is either 0, +1 or −1. If the branch is not part of the loop, then the element is zero (0); if the branch is part of the loop with a positive flow direction, then it is positive one (+1); and if the branch participates to the loop with a negative flow direction, then it is negative one (−1). In the following, the Hardy Cross method and the modelling process of its algorithm are explained by an example.

Example 4.1

Consider a pipe network comprised of nine (J = 9) junctions, thirteen (N = 13) branches and four (M = 4) loops (Figure 4.4). The data provided include: the external inflows (Q_{in}) and outflows (Q_{out}) at all network junctions, and the resistance coefficients (R_i) and pipe diameters (D_i) for all branches. Although not shown explicitly, the resistance coefficients were estimated beforehand from known values of the friction coefficient (f_i), pipe length (L_i), and pipe diameter (D_i), according to Equation 4.4.

In order to initiate the process, discharge values and directions were assumed for all branches in such a way that the continuity principle was maintained at all junctions (Figure 4.4 and Table 4.1). The inflows to the junction were considered as positive and the outflows as negative. Once the discharges were selected, the connectivity matrix for each of the four loops was established, keeping the convention of positive flows following a clockwise movement within the loop (Table 4.2). After that step, all of the required information was available for initiating the iteration process. The convergence criterion for the maximum correction value of ΔQ_l was set equal to 0.001.

The algorithm converged very rapidly, and after seven iterations the correction ΔQ_l was less than 0.001 (Figure 4.5). The computed discharges are shown in Figure 4.4. The algorithm also estimates the velocities (u) and the head drop at each branch (dh). In this example, the connectivity matrix remained unchanged during the iterations.

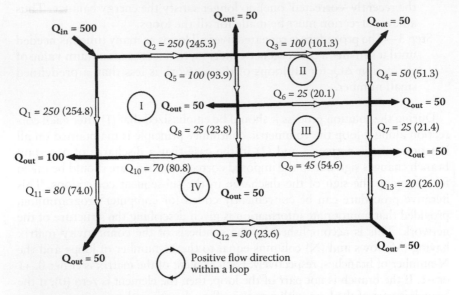

Figure 4.4 Schematic presentation of pipe network. Values in bold were given. Values in italic were assumed. Values in parenthesis were calculated. All flows are in m³/s.

Table 4.1 Input data of discharges, pipe diameters and resistance coefficients

Branch number	Q (m³/s)	D (m)	R (s²/m⁵⁰)
1	250	0.4	96.8
2	250	0.4	96.8
3	100	0.3	306
4	50	0.3	306
5	100	0.3	306
6	25	0.2	3098
7	25	0.2	3098
8	25	0.2	3098
9	45	0.3	306
10	70	0.3	612
11	80	0.3	306
12	30	0.2	9295
13	20	0.2	3098

Table 4.2 Input data of the connectivity matrix

Loop number	Branch number												
	1	*2*	*3*	*4*	*5*	*6*	*7*	*8*	*9*	*10*	*11*	*12*	*13*
1	−1	1	0	0	1	0	0	1	0	−1	0	0	0
2	0	0	1	1	−1	−1	0	0	0	0	0	0	0
3	0	0	0	0	0	1	1	−1	−1	0	0	0	0
4	0	0	0	0	0	0	0	0	1	1	−1	−1	1

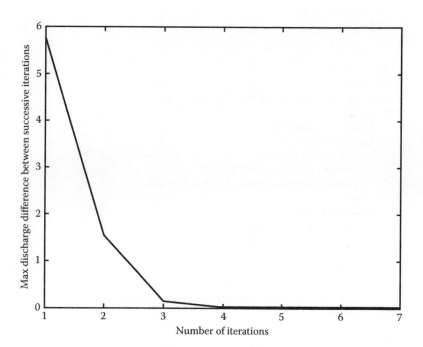

Figure 4.5 Convergence of the Hardy Cross method.

Computer code 4.1

```
% Example 4.1 Pipe Network - Hardy Cross Method
% q = Volumetric discharge [m^3/s];
% d = Pipe diameter [m];
% r = 8*f*l/(g*pi^2d^5) Resistance coefficient [s^2/m^5];
% ctr = Convergence criterion;
% nb = Number of branches;
```

```
% nl = Number of loops;
clc; clear all; close all;
% Input data;
ctr=0.001;
nb=13;
nl=4;
% Import of HCcdr: discharge, pipe diameters, and resistance
coefficients;
load fileqdr.mat
q=HCqdr(:,1);
d=HCqdr(:,2);
r=HCqdr(:,3);
% Import of HCdat: Connectivity matrix [nb x nl];
load filenetstr.mat
mbl=HCdat;
iter=0;
for n=1:500
    iter=iter+1;
    difmax=0;
    % Discharge corrections in each loop;
    for j=1:nl
        sum1=0; sum2=0;
        for k=1:nb
            tt=mbl(j,k);
            sum1=sum1+tt*r(k)*q(k)^2;
            sum2=sum2+2*abs(tt)*r(k)*q(k);
        end
        dq=-sum1/sum2;
        if abs(dq)>difmax;
            difmax=abs(dq);
        end
        for k=1:nb
            q(k)=q(k)+mbl(j,k)*dq;
            % Correction of discharge direction and
connectivity matrix;
            if q(k)<0
                q(k)=-q(k);
            end
            for jj=1:nl
                mbl(jj,k)=-mbl(jj,k);
            end
        end
    end
    it=n;
    diff(it)=difmax;
    % Check for convergence;
    if difmax<ctr
        break
    end
end
```

```
for k=1:nb
    % Flow velocities in the various branches;
    u(k)=q(k)/1000*1.273/d(k)^2;
    % Head loss in the various branches;
    dh(k)=r(k)*q(k)^2/1000000;
end
plot(1:n,diff,'k','Linewidth',1.5)
xlabel('Number of iterations'),ylabel('Max discharge
difference between successive iterations')
```

PROBLEM 4.1

Solve the same problem by making the following suggested changes while keeping the rest of the data constant:

1. Since both the resistance coefficient and the pipe diameter are given, the quantity $fL = \dfrac{g\pi^2}{8}RD^5$ can be estimated for each branch. Based on the data given, find out whether a constant value for the friction coefficient (f) was used. If yes, what was that value?
2. Calculate the discharge in all branches by considering an inflow of 500 m³/s at the junction (1–2) (i.e. connecting branches 1 and 2) and the same amount of outflow at junction (12–13).
3. What kind of changes on pipe diameter would you implement on the given network in order to keep velocity in all branches less than 1.5 m/s?
4. If the network pipes were to be replaced with pipes of the same material but of constant diameter D = 0.2 m, what changes in the computer program would you implement to solve the problem?
5. Solve the same pipe network problem as in Example 4.1 but during your initial selection of discharge values take Q_1 = 450 m³/s and Q_2 = 50 m³/s. The rest of the initial discharges should be selected in such a way so that continuity is preserved at each junction. After how many iteration steps is the correction ΔQ_1 reduced to less than 0.001?

Compare the simulation data obtained by running the modifications, derive conclusions and discuss the significance of the various variables involved to the problem of the pipe network and the Hardy Cross method.

4.3 UNSTEADY FLOW IN A CLOSED CONDUIT

For most practical situations in pipe hydraulics, the water is considered as incompressible and the pipe as rigid. However, those assumptions are not always true. For instance, when there is a sudden change of the flow rate, due to valve closure, pump or turbine start-up or shut-off, and so on, the force caused by the rapid change in flow momentum creates pressures multiple times higher than the hydrostatic pressures. The very high pressures developed, in a phenomenon known as water hammer, result in elastic deformation of both water and pipe. Description of the water hammer phenomenon requires knowledge of the temporal variation of both velocity u(x,t) and pressure head h(x,t) along the pipe. This can be accomplished by introducing the continuity and the momentum equations where the compressibility effects have been accounted for (Sharp and Sharp 1995).

4.3.1 Continuity equation for water hammer

Consider a control volume in an elastic pipe carrying a compressible fluid. The difference between the inflow $Q_{in} = \rho Au$ and outflow $Q_{out} = \rho uA + \dfrac{\partial(\rho uA)}{\partial x}dx$ rates equals the change of the fluid mass within the control volume $\dfrac{\partial(\rho A)dx}{\partial t}$ (Figure 4.6). That yields the continuity equation

$$\frac{\partial(\rho A)}{\partial t} + \frac{\partial(\rho uA)}{\partial x} = 0 \tag{4.15}$$

where ρ is the fluid density, u is the flow velocity and A is the cross-sectional area.

After some minor manipulation Equation 4.15 can be re-written as

$$\frac{1}{\rho}\frac{\partial\rho}{\partial t} + \frac{u}{\rho}\frac{\partial\rho}{\partial x} + \frac{1}{A}\frac{\partial A}{\partial t} + \frac{u}{A}\frac{\partial A}{\partial x} + \frac{\partial u}{\partial x} = 0 \tag{4.16}$$

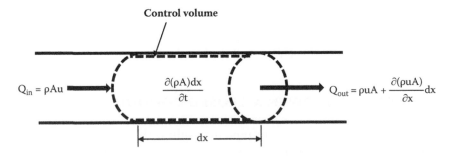

Figure 4.6 Control volume for mass balance analysis.

and by making use of the total (or material derivative) convention, the continuity equation can be written furthermore as

$$\frac{1}{\rho}\frac{d\rho}{dt} + \frac{1}{A}\frac{dA}{dt} + \frac{\partial u}{\partial x} = 0 \tag{4.17}$$

The first term of Equation 4.17 refers to fluid (water) compressibility effects, and the second term describes the pipe's elastic behaviour. Using the definition of the bulk modulus of elasticity of a fluid (E_f), the change of fluid density in time is defined as

$$\frac{1}{\rho}\frac{d\rho}{dt} = \frac{1}{E_f}\frac{dp}{dt} \tag{4.18}$$

where p is the fluid pressure. Assuming an elastic deformation of the pipe due to fluid pressure and neglecting the Poisson ratio, the rate of cross-sectional deformation can be defined as

$$\frac{1}{A}\frac{dA}{dt} = \frac{D}{eE_p}\frac{dp}{dt} \tag{4.19}$$

where D is the pipe diameter, e is the pipe wall thickness and E_p is Young's modulus of elasticity of the pipe material. Combining Equations 4.17 to 4.19 yields

$$\frac{1}{\rho}\frac{dp}{dt} + \left(\frac{\dfrac{E_f}{\rho}}{1 + \dfrac{E_f}{E_p}\dfrac{D}{e}}\right)\frac{\partial u}{\partial x} = \frac{1}{\rho}\frac{dp}{dt} + c^2\frac{\partial u}{\partial x} = 0 \tag{4.20}$$

where c is the speed of elastic wave travelling through the pipe. In terms of the piezometric head (h = p/γ) and the velocity (u), for a horizontal pipe the continuity equation becomes

$$\frac{\partial h}{\partial t} + u\frac{\partial h}{\partial x} + \frac{c^2}{g}\frac{\partial u}{\partial x} = 0 \tag{4.21}$$

Equation 4.21 is a nonlinear partial differential equation of the hyperbolic type. It should be noted that in the case of an inclined pipe, an additional term u sinθ (where θ is the angle of the pipe axis with the horizontal direction) should be added to Equation 4.21, since the piezometric head is h = p/γ + z (see Equation 4.3).

4.3.2 Momentum equation for water hammer

Considering a horizontal pipe, taking the balance of forces acting on a control volume (Figure 4.7) and applying Newton's second law, the momentum equation after some minor manipulation reads

$$\frac{\partial u}{\partial t} + u\frac{\partial u}{\partial x} + \frac{1}{\rho}\frac{\partial p}{\partial x} + \frac{4\tau_o}{\rho D} = 0 \tag{4.22}$$

where τ_o is the wall shear stress and D is the pipe diameter.

In terms of the Darcy-Weisbach equation (Equation 4.4) the wall shear stress (τ_o) can be expressed as

$$\frac{\tau_o}{\rho} = \frac{f}{8}u|u| \tag{4.23}$$

Substituting Equation 4.23 into Equation 4.22 and accounting for hydrostatic pressure yields

$$\frac{\partial u}{\partial t} + u\frac{\partial u}{\partial x} + g\frac{\partial h}{\partial x} + \frac{4K_f}{D}u|u| = 0 \tag{4.24}$$

where $K_f = \frac{f}{8}$. The momentum equation (Equation 4.24) is a nonlinear partial differential equation of the hyperbolic type. Thus the water hammer phenomenon is described in terms of the velocities u(x,t) and pressures h(x,t) by a system of two nonlinear hyperbolic equations (Equations 4.21 and 4.24) that can only be solved by means of numerical analysis.

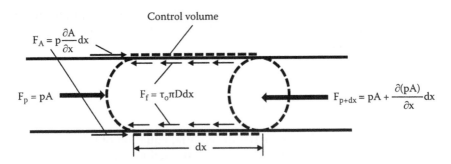

Figure 4.7 Control volume for force balance analysis.

4.3.3 Linearization of water hammer equations

By neglecting advection and frictional effects, the system of Equations 4.21 and 4.24 can be linearized to give

$$\frac{\partial h}{\partial t} + \frac{c^2}{g} \frac{\partial u}{\partial x} = 0 \qquad (4.25)$$

$$\frac{\partial u}{\partial t} + g \frac{\partial h}{\partial x} = 0 \qquad (4.26)$$

Multiplying Equation 4.26 by c^2/g and taking the derivative with respect to x, and by taking the derivative of Equation 4.25 with respect to t, one of the two variables (u, h) can be eliminated, leading to either of the following equations:

$$\frac{\partial^2 h}{\partial t^2} - c^2 \frac{\partial^2 h}{\partial x^2} = 0 \qquad (4.27)$$

$$\frac{\partial^2 u}{\partial t^2} - c^2 \frac{\partial^2 u}{\partial x^2} = 0 \qquad (4.28)$$

Equations 4.27 and 4.28 are linear hyperbolic equations verifying the fact that the 'signals' of either the velocity or the pressure head propagate as waves with speed c, equal to the celerity of elastic waves. For metal ducts carrying water, this celerity is of the order of 10^3 m/s, suggesting that the water hammer phenomenon develops in a matter of a few seconds, a time frame sufficient for the development of high, sometimes destructive, pressures within the pipe.

4.3.4 Numerical treatment of water hammer equations

In most operational applications, the computer simulation model is comprised of the linearized continuity equation and the momentum equation, where advection is neglected but the nonlinear term of wall friction is accounted for. The frictional term does not introduce instabilities but allows the simulation of energy losses, leading to a final equilibrium condition. For the problem to be well-posed, the appropriate initial and boundary conditions are provided.

4.3.4.1 Initial conditions

The initial conditions refer to the pre-existing steady flow, characterized by a constant discharge within the conduit and a linear decrease of the pressure head from a maximum value upstream (in the reservoir feeding the conduit) to a minimum value downstream (zero relative pressure in the case of a free outflow to the atmosphere). By using the Darcy-Weisbach equation, the initial steady state velocity (u_o) is

$$u_o = \sqrt{\frac{gDH_o}{4LK_f}} \qquad (4.29)$$

H_o is the pressure head at the feeding reservoir, and L is the length of the conduit.

4.3.4.2 Boundary conditions

The boundary conditions refer to the pressure and velocity values at both ends of the pipe. Those variables are interrelated in each cross-section via the local specifics (so two conditions are necessary). As explained in the following, the boundary conditions applied are (1) on the downstream section, a known relation between time and discharge (the valve closure schedule); and (2) on the upstream condition (first reach of the conduit inside the reservoir), where pressure head is known (i.e. the water depth of the reservoir). The valve closure schedule $(u_{nx}(t))$ is expressed as

$$u_{nx} = u_o\left(1 - \frac{n\Delta t}{T}\right) \qquad (4.30)$$

where n is the number of time steps passed, Δt is the time step and T is the valve closure time.

4.3.4.3 Numerical algorithm

To avoid the collocated calculation of u and h, we use a staggered computational grid. According to this discretization procedure, the spatial domain is divided into a number of cells. However, the computational points for u and h are not the same, since the velocities are calculated at each side of the cell, while the pressure heads are calculated at the mid-point of the cell (and remain constant throughout the cell). This type of spatial discretization, known as the Arakawa C-grid, ensures the conservation of mass in the numerical solution (Figure 4.8).

Indeed, the scheme equates the difference of incoming and outgoing discharges from the two sides of the cell to the change of the volume of water

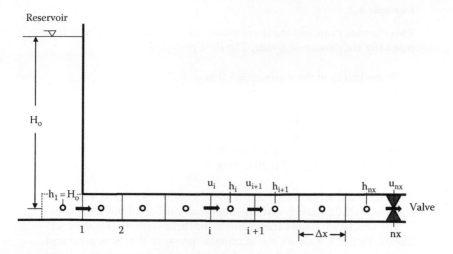

Figure 4.8 Schematic representation of the Arakawa C-grid.

stored within the cell. Using the notation of Figure 4.8, the governing partial differential equations (PDEs) are discretized by using a forward finite differences scheme for the time derivatives and a central scheme for the space derivatives. The resulting numerical equations lead to the following solution scheme:

$$\frac{u_i^{n+1} - u_i^n}{\Delta t} = -g\frac{h_i^n - h_{i-1}^n}{\Delta x} - \frac{4K_f}{D}|u_i^n|u_i^n \tag{4.31}$$

$$\frac{h_i^{n+1} - h_i^n}{\Delta t} = -\frac{c^2}{g}\frac{u_{i+1}^n - u_i^n}{\Delta x} \tag{4.32}$$

The scheme explicitly solves for the velocities and the piezometric heads in a 'leap-frog' manner:

$$u_i^{n+1} = u_i^n - g\frac{\Delta t}{\Delta x}\left(h_i^n - h_{i-1}^n\right) - \frac{4K_f\Delta t}{D}|u_i^n|u_i^n \tag{4.33}$$

$$h_i^{n+1} = h_i^n - \frac{c^2}{g}\frac{\Delta t}{\Delta x}\left(u_i^{n+1} - u_i^n\right) \tag{4.34}$$

The solution arrangement demonstrated by these equations negates the need for solution of an algebraic system of equations for $u(x,t)$ and $h(x,t)$.

Example 4.2

This exercise examines the development of water hammer effects in a pipe after the closure of a vane. The data provided are

Water height in the reservoir = 5.0 m
Elastic wave celerity = 500 m/s
Spatial cell size = 50 m
Number of cells = 100
Pipe diameter = 2 m
Wall friction coefficient (K_f) = 0.05
Valve closure time = 5 s, 20 s, 40 s

The numerical scheme used is given by Equations 4.33 and 4.34. The time step is 0.02 seconds and the simulation ran for 10,000 time steps.

The simulation results are presented for three valve-closing time scenarios. Figure 4.9 shows the maximum pressure that was generated throughout the length of the pipe due to the valve closure. It also shows the assumption made that the surface of the feeding reservoir is unaffected and remains at a height of 5 metres.

Figure 4.10 shows the time history of pressure variations at the end of the pipe (i.e. valve location). The oscillatory behaviour of the elastic wave (bouncing back and forth) and the damping effects of the wall friction are clearly demonstrated.

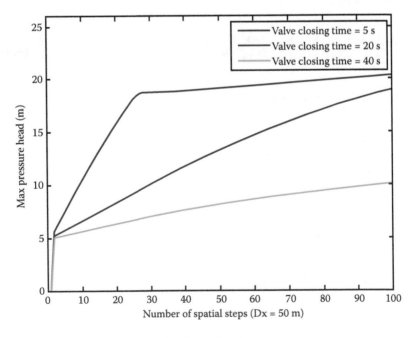

Figure 4.9 Maximum pressure head due to valve closure.

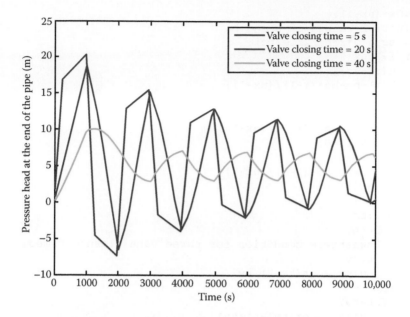

Figure 4.10 Pressure head variations at the end of the pipe.

Computer code 4.2

```
% Example 4.2 Water Hammer Analysis
% r = Pipe radius [m];
% c = Celeriy of the elastic wave [m/s];
% ho = upstream pressure head [m];
% fc = Pipe friction coefficient;
% tv = Valve closing time [s];
% Dt = Time step [s];
% Dx = Spatial step [m];
% nx = Number of discrete spatial points;
% nt = Number of time steps;
% u = Water velocity;
% h = Piezometric head;
clc; clear all; close all;
% Input data;
r=1;
c=500;
ho=5;
fc=0.05;
tv=5; tv1=20; tv2=40;
Dt=0.02;
Dx=50;
nx=100;
tn=10000;
g=9.81;
```

```
le=Dx*(nx-1);
% Initial values of u and h;
uo= sqrt(g*r*ho/2/le/fc);
for i=1:nx
    u(i)=uo;
    h(i)=ho-ho*(i-1)/(nx-1);
    hmax(i)=0;
    u1(i)=uo;
    h1(i)=ho-ho*(i-1)/(nx-1);
    h1max(i)=0;
    u2(i)=uo;
    h2(i)=ho-ho*(i-1)/(nx-1);
    h2max(i)=0;
end
for n=1:tn
    t=Dt*n;
    % Downstream condition for three vane closure times;
    if t<tv
        u(nx)=uo*(tv-t)/tv;
    end
    if t<tv1
        u1(nx)=uo*(tv1-t)/tv1;
    end
    if t<tv2
        u2(nx)=uo*(tv2-t)/tv2;
    end
    % Calculation of new water velocities;
    for k=2:nx-1
        ff=2*fc*u(k)*abs(u(k))/r;
        u(k)=u(k)-Dt/Dx*g*(h(k)-h(k-1))-Dt*ff;
        ff1=2*fc*u1(k)*abs(u1(k))/r;
        u1(k)=u1(k)-Dt/Dx*g*(h1(k)-h1(k-1))-Dt*ff1;
        ff2=2*fc*u2(k)*abs(u2(k))/r;
        u2(k)=u2(k)-Dt/Dx*g*(h2(k)-h2(k-1))-Dt*ff2;
    end
    % Calculation of piezometric head;
    for k=2:nx-1
        h(k)=h(k)-Dt/Dx*c^2/g*(u(k+1)-u(k));
        h1(k)=h1(k)-Dt/Dx*c^2/g*(u1(k+1)-u1(k));
        h2(k)=h2(k)-Dt/Dx*c^2/g*(u2(k+1)-u2(k));
        % Max piezometric head value for each pipe section;
        if h(k)>hmax(k)
            hmax(k)=h(k);
        end
        if h1(k)>h1max(k)
            h1max(k)=h1(k);
        end
        if h2(k)>h2max(k)
            h2max(k)=h2(k);
        end
```

```
              hend(n)=h(nx-1);
              h1end(n)=h1(nx-1);
              h2end(n)=h2(nx-1);
        end
end
hmax(nx)=hmax(nx-1);
h1max(nx)=h1max(nx-1);
h2max(nx)=h2max(nx-1);
plot(1:nx, hmax,'r','Linewidth',1.5)
hold on
plot(1:nx,h1max,'b','Linewidth',1.5)
plot(1:nx, h2max,'g','Linewidth',1.5)
xlabel('Number of spatial steps (Dx = 50m)'), ylabel('Max
pressure head [m]')
legend('valve closure time = 5s','valve closing time = 20s',
'valve closing time = 40s')
axis([0,100,0,26])
figure,plot(1:tn,hend,'r','Linewidth',1.5)
hold on
plot(1:tn,h1end,'b','Linewidth',1.5)
plot(1:tn,h2end,'g','Linewidth',1.5)
xlabel('Time [s]')
ylabel('Pressure head at the end of the pipe [m]')
legend('valve closure time = 5s','valve closing time = 20s',
'valve closing time = 40s')
```

PROBLEM 4.2

Solve the same problem by making the following suggested changes while keeping the rest of the data constant:

1. Sequentially change the elastic wave celerity to 1500 m/s and the pipe diameter to 0.5 m. Run the program and comment on the results.
2. Change the wall friction coefficient to 0.005 and 0.001. Run the program and comment on the results.
3. Neglect the frictional effects, run the simulation and comment on the results.
4. Write a computer program that solves Equation 4.27 by using the explicit scheme described by Equation 3.28 (see Chapter 3). Run and comment on the results.
5. Modify the program as needed by assuming that the pipe is rigid and that only the water is compressible ($E_f = 2.2 \times 10^2$ KN/m^2). Run the program and compare the results with the present example.

Compare the solution data obtained by running the modifications, derive conclusions and discuss the significance of the various parameters involved in the water hammer phenomenon.

Chapter 5

Free surface flows

5.1 QUASI-HORIZONTAL FREE SURFACE FLOWS

Free surface flows are gravity-driven flows that occur in open water systems such as canals, rivers, lakes and oceans. For the majority of practical applications, the study of free surface flows can be simplified by assuming that the curvature of streamlines is small, leading to negligible vertical velocities and accelerations, and consequently to hydrostatic pressure distribution (Figure 5.1). The assumption of this type of 'quasi-horizontal' flows facilitates the mathematical solution since the pressure can be estimated directly from the water depth.

Quasi-horizontal flows are common in rivers and artificial channels, in lakes and reservoirs and in coastal basins. Although the assumption of small-flow curvature may be deviated locally, such as in flow over a weir, the quasi-horizontal conditions are re-established after a short distance and do not affect the far-field conditions. The assumption of quasi-horizontal flow can be extended to bed slopes up to 1/4 even 1/3 with sufficient accuracy (Chow 2009).

5.2 ONE-DIMENSIONAL OPEN CHANNEL FLOW

The main assumptions made for the mathematical treatment of one-dimensional open channel flows are the following:

1. The longitudinal dimension of the flow domain is much larger than the transverse dimensions (width and depth of the channel).
2. The unknown variables are the average cross-sectional velocity $u(x,t)$ and the water depth $y(x,t)$ measured from the free surface to the lowest point of the channel bed. This depth can be expressed either as the sum of the initial depth, $h_o(x)$ and the surface elevation $\zeta(x,t)$ above the initial depth, or the difference between the elevation of the free surface and the elevation of the channel bed, measured from a certain reference datum.

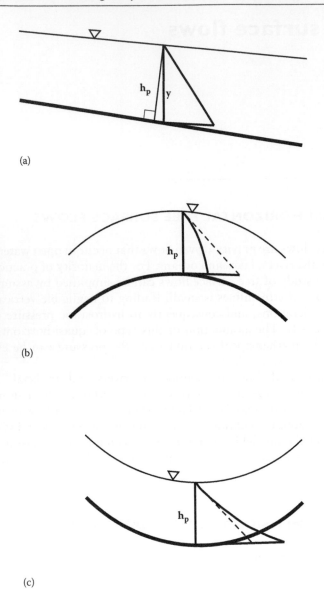

Figure 5.1 (a) Water pressure over linear slope bed. (b) Water pressure over convex-shaped bed. (c) Water pressure over concave-shaped bed.

3. Due to small bed slopes, the hydrostatic pressure is approximated by considering the vertical depth (y), instead of the depth perpendicular to the bed (h_p) (Figure 5.1).
4. The approximation of nearly horizontal flow and that of hydrostatic pressure distribution is valid.

In natural systems, free surface flows are mainly turbulent and thus independent of the Reynolds number (R_e). On the other hand, the behaviour of these flows depends heavily on the Froude number:

$$F_r = \frac{u}{\sqrt{gy}}$$

(5.1)

For $F_r = 1$ flow is critical, for $F_r < 1$ flow is subcritical and for $F_r > 1$ flow is supercritical. The Froude number is indicative of the balance between inertia and gravity forces, and is very important in mathematical modelling since it affects the flow behaviour and the correct selection of the boundary conditions.

Another important classification for open channel flows is their variability in time, t, or distance, x. Thus flows can be steady $\frac{\partial(\varphi)}{\partial t} \equiv 0$ or unsteady $\frac{\partial(\varphi)}{\partial t} \neq 0$, uniform $\left(\frac{\partial(\varphi)}{\partial x} \equiv 0 \right)$ or non-uniform $\frac{\partial(\varphi)}{\partial x} \neq 0$, where φ is either the depth y(x,t) or the velocity u(x,t). Furthermore, non-uniform flows can be gradually varied (surface profiles) or rapidly varied (e.g. surge or bore).

5.2.1 Steady-state gradually varied non-uniform flow

Two important features of any one-dimensional open channel flow are the normal depth, y_n, and the critical depth, y_c. The normal depth corresponds to the water depth under uniform flow conditions and can be estimated using Manning's equation (Equation 5.2) or Chezy's equation (Equation 5.3):

$$u = \frac{Q}{A} = \frac{1}{n} R_h^{2/3} S_o^{1/2}$$

(5.2)

$$u = C_z \sqrt{R_h S_o}$$

(5.3)

where R_h is the hydraulic radius; S_o is the bed slope; and n and C_z are the Manning's and the Chezy's coefficients of friction, respectively. Furthermore, the hydraulic radius is expressed as

$$R_h = \frac{A}{P_w}$$

(5.4)

where A is the flow cross-sectional area and P_w is the wetted perimeter defined as the water–solid boundary interface (Figure 5.2).

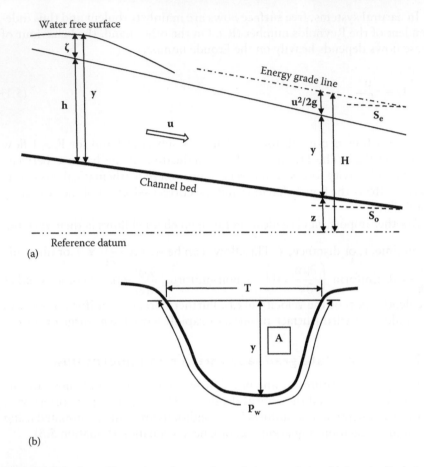

Figure 5.2 Schematic illustration of open channel characteristics. (a) Longitudinal view and (b) cross-sectional area.

It should be noted that in Equations 5.2 and 5.3 the bed slope (S_o) is being used for the estimation of the normal depth, that is, uniform flow. For non-uniform flow, the energy gradient (S_e) should be used instead.

For an arbitrary-shaped cross-section area, the critical depth (y_c) is estimated from the equation

$$\frac{Q^2T}{gA^3} = 1 \tag{5.5}$$

where Q is the discharge and T is the surface flow width. Equation 5.5 is a generalized expression of the Froude number raised to the second power. For rectangular channels, Equation 5.5 can be simplified and solved for the critical depth (y_c) as

$$y_c = \sqrt[3]{\frac{q^2}{g}} \qquad (5.6)$$

where q is the flow discharge per unit width.

5.2.1.1 Newton-Raphson method for estimation of the normal and critical depths

Since both y_n and y_c are implicitly involved in Equations 5.2, 5.3 and 5.5, solutions can be obtained very effectively using a numerical iterative algorithm such as the Newton-Raphson method. By considering a trapezoidal-shaped channel (Figure 5.2), the equations that describe its geometric characteristics are as follows:

$$A = (b + my)y \qquad (5.7)$$

$$P_w = b + 2y\sqrt{1 + m^2} \qquad (5.8)$$

$$T = b + 2my \qquad (5.9)$$

These equations can be also used for rectangular channels (m = 0) or for triangular ones (b = 0). Combining Equations 5.2, 5.7 and 5.8 leads to the following implicit nonlinear function, $fn(y_n)$, for the normal depth:

$$fn(y_n) = \frac{\sqrt{S}}{n} \frac{[(b + my_n)y_n]^{5/3}}{\left[b + 2y_n\sqrt{1 + m^2}\right]^{2/3}} - Q \qquad (5.10)$$

The first derivative of the function Equation 5.10 reads

$$\frac{dfn(y_n)}{dy_n} = fn'(y_n) = \frac{\sqrt{S}}{n}\left[(b + my_n)y_n\right]^{2/3}\left[\left(-\frac{4}{3}\sqrt{1 + m^2}\right)\right.$$

$$\left. + \frac{5}{3}\frac{b + 2my_n}{(b + 2y_n\sqrt{1 + m^2})^{2/3}}\right] \qquad (5.11)$$

Similarly, combining Equations 5.5, 5.7 and 5.9 results into the following implicit non-linear function, fc(y_c), for the critical depth and its first derivative:

$$fc(y_c) = \frac{[(b+my_c)y_c]^{3/2}}{\sqrt{b+2my_c}} - \frac{Q}{\sqrt{g}}$$

(5.12)

$$\frac{dfc(y_c)}{dy_c} = fc'(y_c) = -m\left[\frac{(b+my_c)y_c}{b+2my_c}\right]^{3/2} + \frac{3}{2}\sqrt{[(b+my_c)y_c](b+2my_c)}$$

(5.13)

Once the functions and the derivatives are established, the Newton-Raphson iterative solution algorithm can be applied as

$$y^{(k+1)} = y^{(k)} - \frac{f(y^{(k)})}{f'(y^{(k)})}$$

(5.14)

Starting with a very small initial value for $y^{(0)}$, the algorithm converges very rapidly to the solution for either the y_n or y_c. Estimation of the normal and critical depth is very important for the classification of the water surface profiles for steady-state gradually varied flows.

5.2.1.2 Water surface profiles

The total energy head, H, at any point of the channel is the summation of kinetic energy, potential piezometric energy and potential elevation energy (Figure 5.2):

$$H = \frac{Q^2}{2gA^2} + \frac{p}{\gamma} + z = \frac{Q^2}{2gA^2} + y + z$$

(5.15)

By taking the derivative of Equation 5.15 with respect to x it leads to

$$\frac{dH}{dx} = -\frac{Q^2}{gA^3}\frac{dA}{dx} + \frac{dy}{dx} + \frac{dz}{dx} = -\frac{Q^2T}{gA^3}\frac{dy}{dx} + \frac{dy}{dx} + \frac{dz}{dx}$$

(5.16)

After substitution for the energy gradient, bed slope and Froude number, and by rearranging, an ordinary non-linear differential equation is derived for the free surface profile:

$$\frac{dy}{dx} = \frac{S_o - S_e}{1 - F_r^2}$$

(5.17)

By employing Manning's equation (Equation 5.2) for the energy gradient, S_e, the flow depth, y, is implicitly contained in the variables A, P_w and T.

$$\frac{dy}{dx} = \frac{S_o - n^2 Q^2 \dfrac{P_w^{4/3}}{A^{10/3}}}{1 - \dfrac{Q^2}{g} \dfrac{T}{A^3}}$$

(5.18)

For the solution of Equation 5.18 the appropriate boundary conditions should be provided at some control section (weir, free outflow, etc.). Depending on the relation of the two depths – normal depth, y_n, and critical depth, y_c – the channels are classified as follows:

Mild slope channels (M): $y_c < y_n$
Steep slope channels (S): $y_c > y_n$
Critical slope channels (C): $y_c = y_n$

In addition, by using the slope as a criterion, two more categories are defined:

Horizontal slope channel (H): $S_o = 0$
Adverse slope channel (A): $S_o < 0$

Excluding special cases, the two categories of practical interest are the mild and the steep slope channels. In addition, the surface profiles are categorized based on the water surface location relative to the normal and critical flow depths as follows:

M1 curve: $y > y_n > y_c$
M2 curve: $y_n > y < y_c$
M3 curve: $y_n < y_c < y$
S1 curve: $y > y_c > y_n$
S2 curve: $y_c > y > y_n$
S3 curve: $y_c > y_n > y$

It should be kept in mind that for subcritical flows (curves M1, M2 and S1) the control section (boundary condition) is downstream, thus the solution progresses upstream; whereas for supercritical flows (curves M3, S2 and S3), the control section is upstream and the solution progresses downstream.

The most widely used solution algorithm for Equation 5.18 is the standard step method. However other iterative algorithms for ordinary differential equations (ODEs) can also be applied. For the case of channels with constant cross-sectional area, the second-order Runge-Kutta method can be effectively used as follows. Starting from the control section, the value for Equation 5.18 is calculated as

$$f(y_i) = \frac{S_o - n^2 Q^2 \left(\dfrac{P_w^{4/3}}{A^{10/3}} \right)_{y=y_i}}{1 - \dfrac{Q^2}{g} \left(\dfrac{T}{A^3} \right)_{y=y_i}} \tag{5.19}$$

After selecting a spatial computational step Δx, an adjacent value for the water depth (upstream or downstream) is estimated as

$$y_{i+1}^* = y_i + \frac{1}{2} f(y_i) \Delta x \tag{5.20}$$

Then the function $f^*(y_{i+1})$ is re-evaluated and a corrected value is assigned for the water depth of the adjacent section as

$$y_{i+1} = y_i + f(y_{i+1}^*) \Delta x \tag{5.21}$$

The algorithm continues until $|y_{i+1} - y_i| < \varepsilon$, where ε is a predefined small number.

Example 5.1

This exercise estimates the water surface profiles for a prismatic channel of trapezoidal, rectangular or triangular shape. The main data provided are as follows:

Flow rate = 25 m³/s
Channel length = 4000 m
Longitudinal bed slope = 0.001
Manning's coefficient of friction = 0.025
Bottom width = 5 m (for trapezoidal and rectangular cross-sections)
Side slope = 1:1 (for trapezoidal and triangular cross-sections)

First, the critical and normal depths are calculated by using the Newton-Raphson iterative algorithm (Equation 5.14) with a computational step of 0.1 m. Once those depths are established, the ODE (Equation 5.18) is solved for the water depth $y(x)$ by using the Runge-Kutta method (Equations 5.19 to 5.21). By selecting the appropriate boundary conditions the computer model can accommodate M1, M2,

S2 and S3 surface profiles. More specifically, some suggested boundary conditions are

M1 profile: $Y_d = 1.5y_n$
M2 profile: $Y_d = (y_n + y_c)/2$
S2 profile: $Y_u = (y_n + y_c)/2$
S3 profile: $Y_u = 0.1y_n$

where Y_d and Y_u are the water depths at the downstream and upstream control section, respectively. The computational step was selected as $\Delta x = 1$ m.

For the case of a trapezoidal channel, the computed critical depth ($y_c = 1.25$ m), normal depth ($y_n = 2.19$ m) and the water surface M1 profile are illustrated in Figure 5.3. From that figure it can be easily seen that the M1 profile extends to approximately 2500 m upstream from the control section.

Computer code 5.1

```
% Example 5.1 Water Surface Profiles
% Q = Volumetric discharge [m^3/s];
% S = Bed slope;
% Sf = Energy gradient;
```

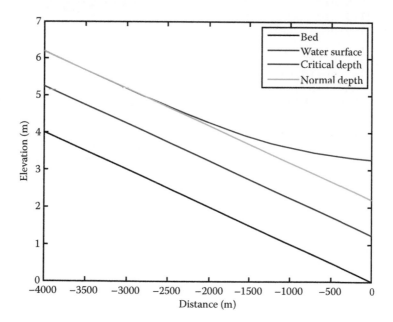

Figure 5.3 Water surface profile (MI curve).

```
% n = Manning's coefficient of friction;
% u = Flow velocity [m/s];
% b = Bottom width [m];
% m = Side slope;
% yn = Normal depth [m];
% yc = Critical depth [m};
% A = Cross sectioanl area {m^2];
% Rh = Hydraulic radius [m];
% Pw = Wetted perimeter [m];
% T = Surface width [m];
% L = Channel lenght [m];
% Y = Water depth [m];
% z= Bed elevation from the reference datum [m];
% h = Water elevation from the reference datum [m];
% Dx = Spatial step [m]:;
clc; clear all; close all;
% Input data;
g=9.81;
Q=25;
S=0.001;
n=0.025;
% Initial depth value to start the iterations;
yin=0.1;
% Trapezoidal, rectangular, triangular cross sections;
% Calculation of the normal depth;
for j=1:3
    b=[5 5 0];
    m=[1 0 1];
    y(1)=yin;
    i=1;
    dyn(1)=0.1;
    while (abs(dyn(i))>0.0001)
        An(i)=b(j)*y(i)+m(j)*(y(i))^2;
        Pwn(i)=b(j)+2*sqrt(m(j)^2+1)*y(i);
        Rhn(i)=An(i)/(Pwn(i));
        Tc(i)=b(j)+2*m(j)*y(i);
        fn(i)=sqrt(S)/n*(An(i)*Rhn(i)^(2/3))-Q;
        dfn(i)=sqrt(S)/n*((-4/3)*Rhn(i)^(5/3)*sqrt(1+m(j)^2)
        +5/3*Rhn(i)^(2/3)*Tc(i));
        y(i+1)=y(i)-fn(i)/dfn(i);
        dyn(i+1)=-fn(i)/dfn(i);
        i=i+1;
    end
    yn(j)=y(i);
    iter_yn(j)=i;
end
% Calculation of the critical depth;
for j=1:3
    yc(1)=yin;
    k=1;
```

```
    dyc(1)=0.1;
    while (abs(dyc(k))>0.0001)
        Ac(k)=b(j)*yc(k)+m(j)*(yc(k))^2;
        Tc(k)=b(j)+2*m(j)*yc(k);
        fc(k)=Ac(k)^(3/2)*Tc(k)^(-1/2)-Q/sqrt(g);
        dfc(k)=-m(j)*(Ac(k)/Tc(k))^(3/2)+(3/2)*sqrt(Ac(k)*T
        c(k));
        %    dfc(k)=-m(j)*Ac(k)^(3/2)*Tc(k)^(-3/2)+
        (3/2)*Tc(k)^(-1/2)*Ac(k)^(1/2)*Tc(k);
        yc(k+1)=yc(k)-fc(k)/dfc(k);
        dyc(k+1)=-fc(k)/dfc(k);
        k=k+1;
    end
    ycr(j)=yc(k);
    iter_yc(j)=k;
end
% Water surface profile calculations for the trapezoidal
channel;
Dx=1;
L=4000;
nx=fix(abs(L/Dx));
% Subcritical flow boundary condition;
% Water depth at the control section yc<yn<y: M1 profile;
Y =1.5*yn(1);
% Water depth at the control section yc<y<yn: M2 profile;
% Y =(yn(1)+ycr(1))/2;
% Supercritical flow boundary condition;
% Water depth at the control section yn<y<yc: S2 profile:
% Y=(yn(1)+ycr(1))/2;
% Water depth at the control section y<yn<yc: S3 profile;
% Y=0.1*yn(1);
if yn-ycr>0
    Dx=-Dx;
end
z(1)=0;
h(1)=Y(1);
b=b(1);
m=m(1);
% Main program;
for i=1:nx;
    A(i)=b*Y(i)+m*Y(i)^2;
    T(i)=b+2*m*Y(i);
    Pw(i)=b+2*Y(i)*sqrt(1+m^2);
    Rh(i)=A(i)/Pw(i);
    u(i)=Q/A(i);
    Sf(i)=n^2*u(i)^2/(Rh(i)^(4/3));
    % ODE f=dy/dx=(S-Sf)/(1-Fr^2);
    f(i)=(S-Sf(i))/(1-Q^2*T(i)/g/A(i)^3);
    yy(i)=Y(i)+0.5*f(i)*Dx;
    A(i)=b*yy(i)+m*yy(i)^2;
```

```
      T(i)=b+2*m*yy(i);
      Pw(i)=b+2*yy(i)*sqrt(1+m^2);
      Rh(i)=A(i)/Pw(i);
      u(i)=Q/A(i);
      Sf(i)=n^2*u(i)^2/(Rh(i)^(4/3));
      fnew(i)=(S-Sf(i))/(1-Q^2*T(i)/g/A(i)^3);
      Y(i+1)=Y(i)+fnew(i)*Dx;
      z(i+1)=-Dx*(i)*S;
      h(i+1)=z(i+1)+Y(i+1);
      x(i+1)=Dx*i;
end
plotyc=z+ycr(1);
plotyn=z+yn(1);
plot(x,z,'k','LineWidth',2); hold on;
plot (x,h,'b','LineWidth',2);
plot(x,plotyc,'r')
plot(x,plotyn,'g')
legend('Bed','Water Surface','Critical Depth','Normal
Depth',1);
xlabel('Distance [m]');
ylabel('Elevation [m]');
title('Water Surface Profile');
```

PROBLEM 5.1

By making the appropriate assumptions, modifying the computer code and changing one or more of the following input data – flow rate, bed slope and Manning's coefficient of friction – as needed, create and plot the water surface profiles under the following scenarios:

1. Subcritical flow with M2 surface profile
2. Supercritical flow with S2 surface profile
3. Supercritical flow with S3 surface profile
4. Supercritical flow with C3 surface profile
5. Subcritical flow with H2 surface profile

Justify and explain your assumptions made and comment on the simulation results.

5.2.2 Unsteady-state open channel flow

5.2.2.1 Governing equations

The unsteady-state one-dimensional quasi-horizontal flows can be described in terms of the average velocity, $u(x,t)$, or the flow discharge, $Q(x,t)$, and the water depth, $y(x,t)$, or the cross-section area, $A(x,t)$. The two governing equations of the phenomenon are based on the principles of mass conservation and momentum balance written along the channel axis, actually along the flow direction (slightly deviating from the horizontal, since the channel slope is usually of the order of less than 1%) (Scarlatos 1996a).

The mass (or volume) conservation (continuity) equation is written as

$$\frac{\partial A}{\partial t} + \frac{\partial Q}{\partial x} = q_L \tag{5.22}$$

where q_L is the lateral inflow. For a rectangular channel Equation 5.22 is written as

$$\frac{\partial y}{\partial t} + \frac{1}{B}\frac{\partial Q}{\partial x} = \frac{q_L}{B} \tag{5.23}$$

Equation 5.23 expresses the evolution of the free surface elevation (or the depth variation) in terms of the difference between the incoming and outgoing discharges through the infinitesimal control space under consideration (the channel reach of length = δx).

For the formulation of the momentum balance equation, the Newton's second law is applied by taking into consideration (Figure 5.4):

- The hydrostatic pressure forces acting on the two sides (upstream and downstream) of the control space
- The component of the weight of water (W_x) within the control volume along the flow direction that is parallel to the bed
- The sum of the distributed frictional forces (τ_o) over the wetted perimeter of the channel

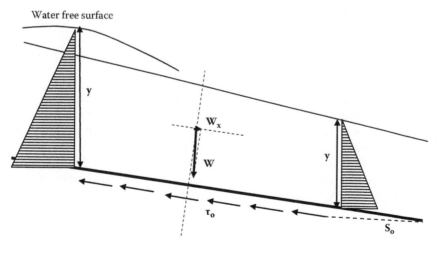

Figure 5.4 Forces acting on the control volume of an open channel.

After some simple manipulation this equation takes the form

$$\frac{\partial u}{\partial t} + u\frac{\partial u}{\partial x} = -g\frac{\partial y}{\partial x} + g(S_o - S_e) \tag{5.24}$$

where S_o is the slope of the channel bed and $S_e = \dfrac{\tau_o}{g\rho R_h}$ is the slope of the energy grade line, a function of the wall friction τ_o. By using the Chezy formula (Equation 5.3) for the wall friction, the operational form for S_e becomes

$$S_e = \frac{u^2}{C_z^2 R_h} \tag{5.25}$$

For rectangular channels, $A = By$ and if the channel width is much bigger than the depth (as in river cross-section geometry), $R_h = y$.

For $q_L = 0$ and by neglecting the nonlinear terms, the governing equations are written as

$$\frac{\partial y}{\partial t} + \frac{1}{B}\frac{\partial Q}{\partial x} = 0 \tag{5.26}$$

$$\frac{\partial u}{\partial t} = -g\frac{\partial y}{\partial x} + g(S_o - S_e) \tag{5.27}$$

In the case of steady flow, the continuity equation is reduced to $Q = $ constant and the equilibrium equation takes the form

$$u\frac{\partial u}{\partial x} + g\frac{\partial y}{\partial x} = g(S_o - S_e) \tag{5.28}$$

which after substitution of $u = Q/A$ leads to the surface profile (Equation 5.17).

The applicability of the unsteady form is very wide, as it describes the propagation of a transient flood wave along a natural stream (e.g. river) or along an artificial canal (e.g. drainage canal).

For a rectangular channel of variable width, the linearized flood propagation model is comprised of the two linear hyperbolic equations of first-order (Equations 5.26 and 5.27), the discharge relation

$$Q = uBy = uA \tag{5.29}$$

and the auxiliary conditions (i.e. initial and boundary).

For the initial conditions, the values of u(x, t = 0) and y(x, t = 0) should be provided for the entire solution domain. In case of an initially 'dry bed', the initial conditions are written as u(x, t = 0) = 0 and y(x, t = 0) = 0.

For subcritical flows boundary conditions should be provided for both the upstream and downstream ends of the channel. For the upstream end, the incoming flood discharge hydrograph Q(x = 0, t) is provided. In general, the discharge hydrograph is described as a discrete-value time series of Q values with an observation time interval δt. For the time interval, hourly values may suffice for a flood event lasting more than a day. In most cases, the observation interval δt is different from the computational time step Δt (see Chapter 3, Figure 3.3).

For the downstream end boundary, either a constant water depth can be maintained (e.g. outflow to a very large water body) or the flow can be led to a critical flow regime (e.g. free outflow) so that no reflected signs can return toward the upstream direction. The first condition is as simple as y(x = L, t) = Y = constant, and the second has the form of a (critical) relation between the flow depth in the last reach and the outflow (Equation 5.6).

5.2.2.2 Numerical solution algorithm

After the discretization of the solution domain, the numerical solution of the system of the governing equations (Equations 5.26, 5.27 and 5.29) is done by means of a staggered grid, similar to that one used for the water hammer phenomenon in closed conduits (Chapter 4, Section 4.3.4.3). Thus, an explicit finite differences scheme based on upwind differences for the time derivatives and central differences for the space derivatives leads to the algebraic equations at point (i,n):

$$\frac{y_i^{n+1} - y_i^n}{\Delta t} = -\frac{2}{B_i + B_{i+1}} \frac{Q_{i+1}^n - Q_i^n}{\Delta x} \tag{5.30}$$

$$\frac{u_i^{n+1} - u_i^n}{\Delta t} = -g \frac{y_i^{n+1} - y_{i-1}^{n+1}}{\Delta x} + g \left(S_0 - \frac{2|u_i^n|u_i^n}{C_z^2(y_i^{n+1} + y_{i-1}^{n+1})} \right) \tag{5.31}$$

$$Q_i^{n+1} = u_i^{n+1} B_i \frac{y_i^{n+1} + y_{i-1}^{n+1}}{2} \tag{5.32}$$

To maintain numerical stability of the solution the Courant-Friedrichs-Lewy criterion must be satisfied at all times (see Chapter 3, Equation 3.29). The upstream boundary condition is the inflow hydrograph $Q_{in}(x = 0, t)$, and the downstream condition is defined by a free flow relationship: $Q_{out} = CBy^{1.5}$.

Example 5.2

This exercise describes the routing of a flood wave over a constant width rectangular channel with an initial dry bed, $y(x, t = 0) = 0$. The data used for the simulation are the following:

Channel width = 100 m
Channel length = 41,000 m
Bed slope = 0.001
Chezy's coefficient of friction = 50 $m^{1/2}/s$

The data for the inflow hydrograph (upstream boundary condition) are shown in Table 5.1. The downstream boundary condition was defined by a flow over weir relationship. The spatial and temporal discretization steps were selected as $\Delta x = 1000$ m and $\Delta t = 5$ s. Since the inflow hydrograph discretization step ($\delta t = 2000$ s) is different from that of the computational time step ($\Delta t = 5$ s) an interpolation procedure is performed.

The simulation was conducted for a period of 45,000 seconds (12.5 hours). Figure 5.5 shows a graphical depiction of the inflow

Table 5.1 Inflow hydrograph

Time (seconds)	0	2000	4000	6000	8000	10,000	12,000	14,000
Discharge (m³/s)	0	0	600	400	200	100	50	0

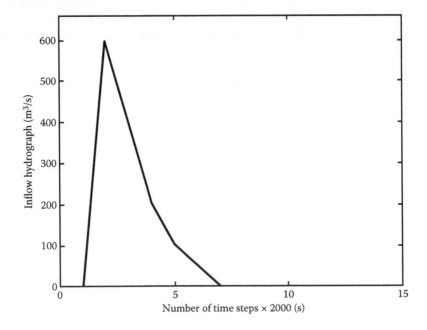

Figure 5.5 Inflow hydrograph.

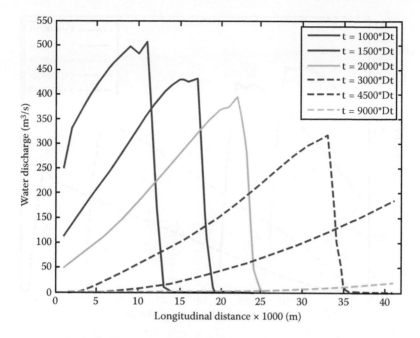

Figure 5.6 **Flow rates over the channel at different times.**

hydrograph. In order to illustrate the flood wave propagation, the flow discharges (Q) and the water depths (h) are plotted over the entire channel at times of 5000, 7500, 10,000, 15,000, 22,500 and 45,000 seconds (Figures 5.6 and 5.7, respectively). These figures show how the peak flow rate and the water depth attenuate in time. Also, the drying of the channel after the passage of the flood waters is evident.

Computer code 5.2

```
% Example 5.2 Unsteady Flow in Rectangular Open Channel of
Constant Width
% h = Flow depth [m];
% b = Channel width [m];
% S = Bed slope;
% C = Chezy coefficient of friction [m^(1/2)/s];
% Qin = Inflow hydrograph [m^3/s];
% Dx = Spatial step [m];
% Dt = Time step [s];
% nx = Number of spatial steps;
% nt = Number of time steps;
% dt = Time step of the inflow hydrograph [s];
% nth = Number of time steps of the inflow hydrograph;
```

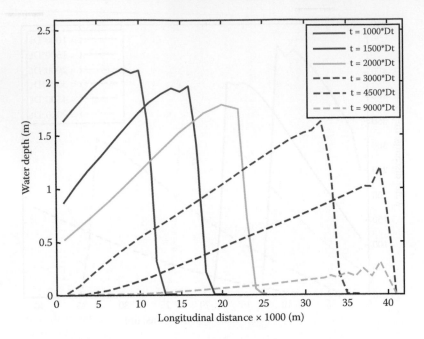

Figure 5.7 Water depths over the channel at different times.

```
clc; clear all; close all;
% Input data;
g=9.81;
Dx=1000;
Dt=5;
nx=41;
nt=9000;
dt=2000;
nth=15;
for i=1:nx
    h(i)=0;
    S(i)=0.001;
    b(i)=100;
    C(i)=50;
end
Qin=[0,600,400,200,100,50,0,0,0,0,0,0,0,0,0];
k=0;
% Main program;
for k=1:nt
    k=k+1;
    % Calculation of inflow discharge by interpolation;
    jj=k*Dt/dt;
    jr=round(jj)+1;
    if jr >= nth
```

```
        jr=nth-1;
end
Q(1)=Qin(jr)+(Qin(jr+1)-Qin(jr))*(k*Dt-(jr-1)*dt)/dt;
% Calculation of the velocity and discharge;
for i=2:nx-1
    if h(i-1)<0.001
        u(i)=0;
    else u(i)=u(i)+Dt*g*(S(i)-(h(i)-h(i-1))/
    Dx-u(i)*abs(u(i))/C(i)^2*2/(h(i)+h(i-1)));
    end
    Q(i)=u(i)*b(i)*(h(i)+h(i-1))/2;
end
% Downstream boundary condition;
Q(nx)=b(nx)*h(nx-1)*sqrt(g*h(nx-1));
% Estimation of the water depths:
for i=1:nx-1
    h(i)=h(i)-Dt/Dx/(b(i)+b(i+1))*2*(Q(i+1)-Q(i));
end
% Data stored at different time intervals for plotting;
if k == round(nt/9)
    for m=1:nx
        h1t(m)=h(m);
        Q1t(m)=Q(m);
    end
end
if k == round(nt/6)
    for m=1:nx
        h2t(m)=h(m);
        Q2t(m)=Q(m);
    end
end
if k == round(nt/4.5)
    for m=1:nx
        h3t(m)=h(m);
        Q3t(m)=Q(m);
    end
end
if k == round(nt/3)
    for m=1:nx
        h4t(m)=h(m);
        Q4t(m)=Q(m);
    end
end
if k == round(nt/2)
    for m=1:nx
        h5t(m)=h(m);
        Q5t(m)=Q(m);
    end
end
if k == nt
```

```
            for m=1:nx
                h6t(m)=h(m);
                Q6t(m)=Q(m);
            end
        end
    end
end
plot(1:nth,Qin,'k','Linewidth',1.5)
xlabel('Number of time steps x 2000 [s]')
ylabel('Inflow hydrograph [m^3/s]')
axis([0,15,0,660])
figure, plot(1:nx,Q1t,'b','Linewidth',1.5)
hold on
plot(1:nx,Q2t,'r','Linewidth',1.5)
plot(1:nx,Q3t,'g','Linewidth',1.5)
plot(1:nx,Q4t,'b--','Linewidth',1.5)
plot(1:nx,Q5t,'r--','Linewidth',1.5)
plot(1:nx,Q6t,'g--','Linewidth',1.5)
xlabel('Longitudinal distance x 1000 [m]')
ylabel('Water discharge [m^3/s]')
axis([0,42,0,550])
legend('t=1000*Dt','t=1500*Dt','t=2000*Dt','t=3000*Dt','t=45
00*Dt','t=9000*Dt')
figure, plot(1:nx,h1t,'b','Linewidth',1.5)
hold on
plot(1:nx,h2t,'r','Linewidth',1.5)
plot(1:nx,h3t,'g','Linewidth',1.5)
plot(1:nx,h4t,'b--','Linewidth',1.5)
plot(1:nx,h5t,'r--','Linewidth',1.5)
plot(1:nx,h6t,'g--','Linewidth',1.5)
xlabel('Longitudinal distance x 1000 [m]')
ylabel('Water depth [m]')
axis([0,42,0,2.6])
legend('t=1000*Dt','t=1500*Dt','t=2000*Dt','t=3000*Dt','t=45
00*Dt','t=9000*Dt')
```

PROBLEM 5.2

Solve the same problem by making the following suggested changes while keeping the rest of the data constant:

1. Change the roughness coefficient linearly from $C_z = 105$ m$^{1/2}$/s at the upstream boundary to $C_z = 30$ m$^{1/2}$/s at the downstream boundary. Repeat the exercise by inverting the direction of roughness change and compare the results.
2. For the upstream half length of the channel, set the slope to $S_0 = 0.001$ and for the downstream half to $S_0 = 0.01$. Check if there would be a transition from subcritical to supercritical flow, run the program and comment on the results.

3. Change the inflow hydrograph to Q = 0 m³/s at t = 0 s; Q = 1000 m³/s at t = 2000 s; Q = 0 m³/s at t = 4000 s; Q = 1000 m³/s at t = 6000 s; and Q = 0 m³/s at t = 8000 s. Assume a linear variation of the hydrographs between the time intervals δt = 2000 s. Run the program and comment on the results.
4. By using the appropriate expression, change the downstream boundary to a zero-flux (reflecting) boundary. Comment on the results.
5. Change the computational time step to Δt = 400 s. Comment on the results.

Example 5.3

This exercise investigates the routing of a sudden-release (dam-break) flood wave in a natural stream of variable width. The data used for the simulation are as follows:

Channel length = 46,000 m
Bed slope = 0.0004
Chezy's coefficient of friction = 40 m$^{1/2}$/s

The data of the variable stream width are shown in Table 5.2. The initial conditions were set as $u(x, t = 0) = 0$ and $h(x, t = 0) = 0.1$ m. The

Table 5.2 Width of the different stream longitudinal segments

Segment number	1	2	3	4	5	6	7	8	9	10
Channel width (m)	1810	1668	1346	1706	1732	1368	1205	1348	1218	1256
Segment number	11	12	13	14	15	16	17	18	19	20
Channel width (m)	1394	1237	1502	1076	1251	1317	1525	1567	1756	282
Segment number	21	22	23	24	25	26	27	28	29	30
Channel width (m)	282	1770	1348	1200	1436	140	1298	1357	1380	636
Segment number	31	32	33	34	35	36	37	38	39	40
Channel width (m)	130	1416	1384	1386	1369	1356	170	1371	260	1374
Segment number	41	42	43	44	45	46				
Channel width (m)	1346	1446	1542	1628	1698	1804				

upstream boundary condition was the inflow hydrograph Q_{in} (m³/s) defined as

$$Q_{in} = 2000\left(1 - \frac{t}{5 \times 10^4}\right) \text{ for } t \le 5 \times 10^4 \text{s}$$

and

$$Q_{in} = 0 \text{ for } t > 5 \times 10^4 \text{s}$$

The downstream boundary condition was defined as free flow. The spatial and temporal discretization steps were selected as $\Delta x = 1000$ m and $\Delta t = 5$ s.

The computation was conducted for a period T_d of 250,000 seconds (69,44 hours) simulating the dam-break water release duration. Figure 5.8 shows a graphical depiction of the dam-break inflow hydrograph. In order to illustrate the flood wave propagation, the flow discharges (Q) and the water depths (h) are plotted over the entire stream at times of $T_d/25$, $T_d/5$ and T_d (Figures 5.9 and 5.10, respectively). These figures show how the peak flow rate and the water depth attenuate in time. Also, the drying of the channel after the passage of the flood waters is evident.

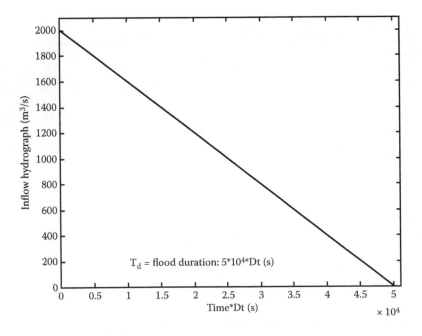

Figure 5.8 Inflow hydrograph.

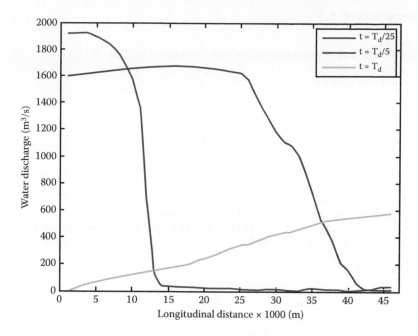

Figure 5.9 Flow rates along the stream at different times.

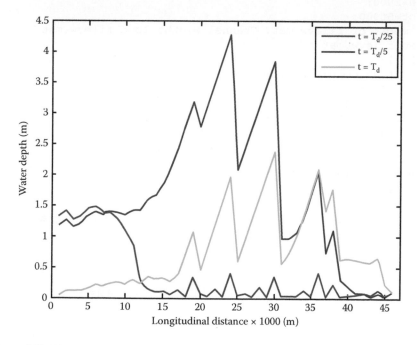

Figure 5.10 Water depths along the stream at different times.

Computer code 5.3

```
% Example 5.3 Unsteady Flow in Rectangular Open Channel of
Variable Width
% h = Flow depth [m];
% B = Channel width [m];
% S = Bed slope;
% C = Chezy coefficient of friction [m^(1/2)/s];
% Qin = Inflow hydrograph [m^3/s];
% Dx = Spatial step [m];
% Dt = Time step [s];
% nx = Number of spatial steps;
% nt = Number of time steps;
% dt = Time step of the inflow hydrograph [s];
% ntot = Number of time steps of the inflow hydrograph;
clc; clear all; close all;
% Input data;
g=9.81;
S=0.0004;
C=40;
Dx=1000;
Dt=5;
nx=46;
nt=50000;
Qin=2000;
ntot=50000;
% Import of filevw.mat: Channel widths;load filevw.mat
load filevw.mat
b=B(:);
% Initial conditions;
for i=1:nx
    h(i)=0.1;
    hmax(i)=0;
    u(i)=0;
end
k=0;
% Main program;
for k=1:nt
    k=k+1;
    % Calculation of linearly changing inflow hydrograph;
    Q(1)=Qin-Qin*k/ntot;
    if k-ntot>0
        Q(1)=0;
    end
    % Calculation of the velocity and discharge;
    for i=2:nx-1
        if h(i-1)-0.02<0
            u(i)=0;
```

```
        else u(i)=u(i)+Dt*g*(S-(h(i)-h(i-1))/
        Dx-u(i)*abs(u(i))/C^2*2/(h(i)+h(i-1)));
        end
        Q(i)=u(i)*b(i)*(h(i)+h(i-1))/2;
    end
    % Downstream boundary condition;
    Q(nx)=b(nx)*h(nx-1)*sqrt(g*h(nx-1));
    % Estimation of the water depths:
    for i=1:nx-1
        h(i)=h(i)-Dt/Dx/(b(i)+b(i+1))*2*(Q(i+1)-Q(i));
    end
    % Data stored at different time intervals for plotting;
    if k == round(nt/25)
        for m=1:nx
            h1t(m)=h(m);
            Q1t(m)=Q(m);
        end
    end
    if k == round(nt/5)
        for m=1:nx
            h3t(m)=h(m);
            Q3t(m)=Q(m);
        end
    end
    if k == round(nt)
        for m=1:nx
            h5t(m)=h(m);
            Q5t(m)=Q(m);
        end
    end
end
for j=1:ntot
    Qinflow(j)=Qin-Qin*j/ntot;
end
plot(1:ntot,Qinflow,'k','Linewidth',1.5)
xlabel('Time*Dt [s]')
ylabel('Inflow hydrograph [m^3/s]')
axis([0,51000,0,2100])
text(10000,200, 'Td=Flood duration: 5*10^4*Dt [s]')
figure, plot(1:nx,Q1t,'b','Linewidth',1.5)
hold on
plot(1:nx,Q3t,'r','Linewidth',1.5)
plot(1:nx,Q5t,'g','Linewidth',1.5)
xlabel('Longitudinal distance x 1000 [m]')
ylabel('Water discharge [m^3/s]')
axis([0,47,0,2000])
legend('t=Td/25','t=Td/5','t=Td')
figure, plot(1:nx,h1t,'b','Linewidth',1.5)
hold on
```

```
plot(1:nx,h3t,'r','Linewidth',1.5)
plot(1:nx,h5t,'g','Linewidth',1.5)
xlabel('Longitudinal distance x 1000 [m]')
ylabel('Water depth [m]')
axis([0,47,0,4.5])
legend('t=Td/25','t=Td/5','t=Td')
```

PROBLEM 5.3

Solve the same problem by making the following suggested changes while keeping the rest of the data constant:

1. Change the width linearly from 100 m at the upstream end to 2000 m at the downstream boundary. Then set a 2000 m upstream width, linearly reducing to 100 m at the downstream end. Run the program and compare the results of the two simulations.
2. Change the inflow hydrograph to a constant Q = 2000 m³/s lasting from t = 0 to t = 3600 seconds. Run the program and comment on the simulation results.
3. By using the appropriate expression, change the downstream boundary to a zero-flux (reflecting) boundary. Run the program and comment on the simulation results.
4. Modify the code as needed to simulate a constant base flow of 1000 m³/s. (Hint: Set the upstream boundary hydrograph at a constant inflow of 1000 m³/s until steady-steady conditions are established.)
5. For a constant flow of 1000 m³/s, make the appropriate changes to the downstream boundary condition to accommodate tidal disturbances. (Hint: Set $Q_{out} = Q_{end} + Q_T sin(\sigma t)$ where Q_{out} is the total outflow, Q_{end} is the base flow discharge, Q_T is the tidal flux and σ is the tidal frequency.) Select values for Q_T and σ, run the simulation for 10 tidal cycles and comment on the results.

5.3 TWO-DIMENSIONAL HORIZONTAL FREE SURFACE FLOWS

For two-dimensional free surface flows, the solution domain is usually a large water basin (e.g. open sea, lake or reservoir) characterised by low water velocities. If the utilized coordinate system is not an inertial one but one following the radial acceleration of earth, the effect of the rotation of the earth, whenever important, appears indirectly. Thus, in the momentum equations for a non-inertial system, Coriolis effects are introduced in the form of an internal distributed force expressed as

$$\vec{C} = -2\rho\vec{\Omega} \times \vec{U} \tag{5.33}$$

where $\vec{\Omega}$, is the vector of the earth radial velocity, $\|\vec{\Omega}\| = \dfrac{2\pi}{86400}$ rad/sec, and $\vec{U} = u\hat{i} + v\hat{j}$ is the fluid particle velocity vector. The components of the Coriolis force in the horizontal flow plane (x–y) are given as

$$\frac{C_x}{\rho} = fv \quad \text{and} \quad \frac{C_y}{\rho} = -fu \tag{5.34}$$

where $f = 2\Omega\sin\varphi$ and φ is the latitude of the geographic location of the flow.

In the case of two-dimensional flows, the unknown variables are the free surface deviation $\zeta(x,y,t)$ from the horizontal still water level $h(x,y)$ and the depth mean horizontal velocity components $u(x,y,t)$ and $v(x,y,t)$. The governing equations are formulated on the basis of the principles of mass conservation and momentum balance along the horizontal x- and y-axes. The external forces acting on the control volume (a water column extending from the bed to the surface with base dimensions Δx, Δy) are the wind friction on the water surface (τ_{sx}, τ_{sy}), and the bottom friction on the solid bed (τ_{bx}, τ_{by}). These shear stresses are commonly expressed as quadratic functions of the velocities: of the wind ($\vec{w} = w_x\hat{i} + w_y\hat{j}$) on the surface and of the water ($\vec{U}$) on the bed, along with the corresponding friction coefficients f_s and f_b:

$$\frac{\tau_{sx}}{\rho} = f_s w_x \sqrt{w_x^2 + w_y^2} \quad \text{and} \quad \frac{\tau_{sy}}{\rho} = f_s w_y \sqrt{w_x^2 + w_y^2} \tag{5.35}$$

$$\frac{\tau_{bx}}{\rho} = f_b u \sqrt{u^2 + v^2} \quad \text{and} \quad \frac{\tau_{by}}{\rho} = f_b v \sqrt{u^2 + v^2} \tag{5.36}$$

The order of magnitude of these dimensionless friction coefficients is $O(f_b) = 10^{-3}$, $O(f_s) = 10^{-6}$. Apart from those external forces, internal frictional forces intervene in the equilibrium, resulting in the diffusion of the momentum in the water body. Since the Reynolds numbers are high, the flows are turbulent and the internal friction is regulated, not by the molecular viscosity, μ, but by the 'eddy' or turbulent viscosity, ε. The internal friction effects can be described, in the simpler possible form, by means of diffusive terms:

$$D_x = \varepsilon_h \left(\frac{\partial^2 u}{\partial x^2} + \frac{\partial^2 u}{\partial y^2} \right) \quad \text{and} \quad D_y = \varepsilon_h \left(\frac{\partial^2 v}{\partial x^2} + \frac{\partial^2 v}{\partial y^2} \right) \tag{5.37}$$

where ε_h is the horizontal eddy viscosity coefficient (Rodi 2000).

5.3.1 Governing equations

The mass conservation and momentum balance equations for a two-dimensional system (Figure 5.11) can be written as follows:

$$\frac{\partial \zeta}{\partial t} + \frac{\partial (hu)}{\partial x} + \frac{\partial (hv)}{\partial y} = 0 \tag{5.38}$$

$$\frac{\partial u}{\partial t} + u \frac{\partial u}{\partial x} + v \frac{\partial u}{\partial y} = -g \frac{\partial \zeta}{\partial x} + fv + \varepsilon_h \left(\frac{\partial^2 u}{\partial x^2} + \frac{\partial^2 u}{\partial y^2} \right) + \frac{\tau_{sx}}{\rho h} - \frac{\tau_{bx}}{\rho h} \tag{5.39}$$

$$\frac{\partial v}{\partial t} + u \frac{\partial v}{\partial x} + v \frac{\partial v}{\partial y} = -g \frac{\partial \zeta}{\partial y} - fu + \varepsilon_h \left(\frac{\partial^2 v}{\partial x^2} + \frac{\partial^2 v}{\partial y^2} \right) + \frac{\tau_{sy}}{\rho h} - \frac{\tau_{by}}{\rho h} \tag{5.40}$$

In these equations the water depth $y(x,y,t)$ was replaced by the undisturbed water depth $h(x,y)$ under the assumption that $\zeta \ll h$. Due to the complexity of Equations 5.38 to 5.40, the system can only be solved numerically.

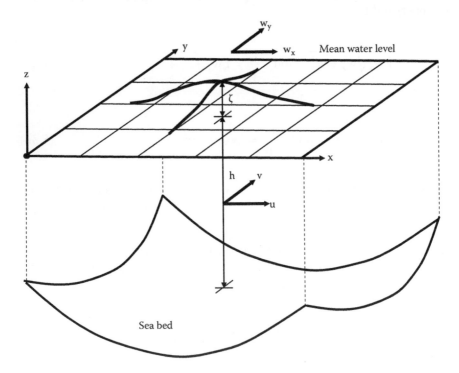

Figure 5.11 Two-dimensional solution domain.

The non-linear terms of the total (material) derivatives Du/Dt and Dv/Dt tend to produce numerical instabilities and chaotic solutions. The frictional terms tend to stabilize the flow and the diffusive terms tend to smooth the functional forms and diminish the development of numerical instabilities. Notably, the equations contain both hyperbolic and parabolic terms.

The selection of the spatial-temporal distribution and magnitude of the horizontal eddy viscosity coefficient, ε_h, is a concurrent problem of the 'turbulence closure'. That coefficient can be used as a controller of the numerical stability and its influence can be merged with the numerical diffusion error. The coefficient ε_h can be properly adjusted either to permit the description of eddies of certain geometric and energy scale, or to diffuse them letting only more basic features of the flow appear in the solution.

For geophysical scale flows, a simple but effective approach to the problem is the use of the Smagorinsky equation, in which the eddy diffusivity–viscosity is related to the gradients of the horizontal velocity components as

$$\varepsilon_h = C(\Delta x)^2 \sqrt{\left(\frac{\partial u}{\partial x} - \frac{\partial v}{\partial y}\right)^2 + \left(\frac{\partial u}{\partial y} + \frac{\partial v}{\partial x}\right)^2} \tag{5.41}$$

where $0.1 \leq C \leq 1.0$. The proper choice of C Equation 5.41 depends on the numerical scheme selected for the time-integration of the governing equations (Equations 5.38 to 5.40).

5.3.2 Initial and boundary conditions

For completeness, the appropriate initial and boundary conditions should be provided. The initial conditions are usually given in the form of a 'cold start' that assigns zero values for the dependent variables ζ and u, v at t = 0.

The boundary conditions involve the description of the friction on the free surface and on the bed, as well as the conditions on the lateral boundaries. The coastal boundaries can be described either by a full suppression of the water velocity (u = v = 0); or by the free-slip condition, where along the boundary only the velocity component normal to the boundary is suppressed.

Another important boundary condition appears in the case of a semi-enclosed geophysical basin (e.g. bay) connected to a huge body of water absorbing, without back reflection, any water surface perturbation signal arriving from the bay. This boundary is called open-sea boundary (OSB) and the boundary condition used must be able to describe

- The incidence from the open sea of known predetermined perturbations, for example, in the case of tidal sea, the sinusoidal variations of the free surface
- The radiation, without return to the open sea, of all signals arriving on the boundary from the inner bay

Those conditions, known as open-sea boundary conditions in the simplest linear form, can be expressed by the sum of two components:

$$\zeta_{total} = \zeta_{incident} + \zeta_{radiated} = \zeta_0 \sin (\omega t) + \zeta_{radiated} \quad (5.42)$$

The radiated term is controlled by the Sommerfeld equation,

$$\frac{\partial \zeta_{radiated}}{\partial t} + c_o \frac{\partial \zeta_{radiated}}{\partial n} = 0 \quad (5.43)$$

where c_o is the celerity of the long wave and n is the outward direction normal to the boundary.

An equivalent condition involving the velocity component normal to the boundary is given as

$$U_n = \pm \zeta \sqrt{\frac{g}{h}} \quad (5.44)$$

where ζ is measured at the cell next to the U_n value. Whenever possible, selection of an OSB line parallel to the x- or y-axis simplifies the computational algorithm.

5.3.3 Radiation stresses

In coastal areas propagating gravity water waves may get modulated due to secondary phenomena like the wave breaking in the surf zone. In that case, a special flow-generating phenomenon appears, due to the variation of the average over the wave period momentum along the wave propagation domain. This phenomenon is described by the three components of the so-called radiation stresses. These radiation stresses – S_{xx}, S_{yy} and $S_{xy} = S_{yx}$ – are components of a second-order tensor. In the case of long waves, the radiation stresses can be described in terms of the depth averaged water velocity components u, v and the free surface elevation ζ averaged over the wave period.

For the simple case of a long, small amplitude linear wave, the water velocity components are related to the free surface elevation ζ through the linear equations

$$\frac{\partial u}{\partial t} = -g \frac{\partial \zeta}{\partial x} = -\frac{c_o^2}{h} \frac{\partial \zeta}{\partial x} \quad (5.45)$$

$$\frac{\partial v}{\partial t} = -g \frac{\partial \zeta}{\partial y} = -\frac{c_o^2}{h} \frac{\partial \zeta}{\partial y} \quad (5.46)$$

and the three components of the radiation stresses are approximated by the expressions

$$\frac{S_{xx}}{\rho} = \frac{1}{T} \int_t^{t+T} \left[hu^2 + \frac{g\zeta^2}{2} - \frac{h}{3}\left(\frac{d\zeta}{dt}\right)^2 \right] dt \qquad (5.47)$$

$$\frac{S_{yy}}{\rho} = \frac{1}{T} \int_t^{t+T} \left[hv^2 + \frac{g\zeta^2}{2} - \frac{h}{3}\left(\frac{d\zeta}{dt}\right)^2 \right] dt \qquad (5.48)$$

$$\frac{S_{xy}}{\rho} = \frac{S_{yx}}{\rho} = \frac{1}{T} \int_t^{t+T} huv \, dt \qquad (5.49)$$

where T is the wave period.

The wave-generated currents terms appear on the right-hand side of the momentum equilibrium equations in the form

$$\frac{\partial u}{\partial t} + u\frac{\partial u}{\partial x} + v\frac{\partial u}{\partial y} = -g\frac{\partial \zeta}{\partial x} + fv + \varepsilon_h\left(\frac{\partial^2 u}{\partial x^2} + \frac{\partial^2 u}{\partial y^2}\right)$$
$$+ \frac{\tau_{sx}}{\rho h} - \frac{\tau_{bx}}{\rho h} - \frac{1}{\rho h}\left(\frac{\partial S_{xx}}{\partial x} + \frac{\partial S_{xy}}{\partial y}\right) \qquad (5.50)$$

$$\frac{\partial v}{\partial t} + u\frac{\partial v}{\partial x} + v\frac{\partial v}{\partial y} = -g\frac{\partial \zeta}{\partial y} - fu + \varepsilon_h\left(\frac{\partial^2 v}{\partial x^2} + \frac{\partial^2 v}{\partial y^2}\right)$$
$$+ \frac{\tau_{sy}}{\rho h} - \frac{\tau_{by}}{\rho h} - \frac{1}{\rho h}\left(\frac{\partial S_{yx}}{\partial x} + \frac{\partial S_{yy}}{\partial y}\right) \qquad (5.51)$$

5.3.4 Numerical solution scheme

The numerical solution of the governing equations is based on a simple explicit and stable scheme, with minimal numerical errors, and easily programmable. The flow domain is discretized by means of a staggered grid

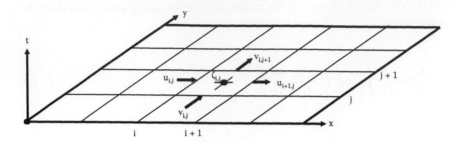

Figure 5.12 Computational grid for estimation of the flow velocities and water depths.

presented in Figure 5.12. The velocity components are computed on the sides of a cell, and the depths and free surface elevations on the centre of the mesh. Thus, the computational algorithm is organized as follows:

1. From the known (available) ζ, u, and v, values at time level n, the new values of ζ at time level n + 1 are computed using the mass conservation equation.
2. From the newly computed values of ζ, the values of u, v, at time n + 1, are computed using the two momentum equilibrium equations.
3. The proximity to lateral solid boundaries or open-sea boundaries is taken into consideration and proper boundary conditions are applied.
4. Once all of the ζ, u, v values are computed at time level n + 1, then the solution proceeds to the computation of the variables at the new time level n + 2.
5. The computed values are periodically stored and the computation returns to step 1, and the process is repeated until the time index reaches a predetermined value of computational time steps.

If the forcing factors of the flow are constant or periodic, the flow variables also become steady or periodic after some time steps, and the stored values provide operational information for subsequent use. According to the notations of Figure 5.12, the numerical approximation of the flow equations is

$$\frac{\zeta_{i,j}^{n+1} - \zeta_{i,j}^{n}}{\Delta t} = -\left[\frac{\left(h_{i+1,j}^{n} + h_{i,j}^{n}\right)u_{i+1,j}^{n} - \left(h_{i,j}^{n} + h_{i-1,j}^{n}\right)u_{i,j}^{n}}{2\Delta x} + \frac{\left(h_{i,j+1}^{n} + h_{i,j}^{n}\right)v_{i,j+1}^{n} - \left(h_{i,j}^{n} + h_{i,j-1}^{n}\right)v_{i,j}^{n}}{2\Delta y} \right] \tag{5.52}$$

$$\frac{u_{i,j}^{n+1} - u_{i,j}^{n}}{\Delta t} = -\frac{u_{i,j}^{n}(u_{i+1,j}^{n} - u_{i-1,j}^{n})}{2\Delta x} - \frac{v_m(u_{i,j+1}^{n} - u_{i,j-1}^{n})}{2\Delta y} - g\frac{\zeta_{i,j}^{n} - \zeta_{i-1,j}^{n}}{\Delta x} + fv_m$$

$$+\varepsilon_h\left(\frac{u_{i+1,j}^{n} + u_{i-1,j}^{n} - 2u_{i,j}^{n}}{(\Delta x)^2} + \frac{u_{i,j+1}^{n} + u_{i,j-1}^{n} - 2u_{i,j}^{n}}{(\Delta y)^2}\right) + \frac{\tau_{sxi,j}}{\rho h_{mx}} - \frac{\tau_{bxi,j}}{\rho h_{mx}}$$

$$(5.53)$$

$$\frac{v_{i,j}^{n+1} - v_{i,j}^{n}}{\Delta t} = -\frac{u_m\left(v_{i+1,j}^{n} - v_{i-1,j}^{n}\right)}{2\Delta x} - \frac{v_{i,j}^{n}\left(v_{i,j+1}^{n} - v_{i,j-1}^{n}\right)}{2\Delta y} - g\frac{\zeta_{i,j}^{n} - \zeta_{i,j-1}^{n}}{\Delta y} - fu_m$$

$$+\varepsilon_h\left(\frac{v_{i+1,j}^{n} + v_{i-1,j}^{n} - 2v_{i,j}^{n}}{(\Delta x)^2} + \frac{v_{i,j+1}^{n} + v_{i,j-1}^{n} - 2v_{i,j}^{n}}{(\Delta y)^2}\right) + \frac{\tau_{syi,j}}{\rho h_{my}} - \frac{\tau_{byi,j}}{\rho h_{my}}$$

$$(5.54)$$

where

$$u_m = \frac{u_{i+1,j}^{n} + u_{i,j}^{n} + u_{i+1,j-1}^{n} + u_{i,j-1}^{n}}{4} \qquad (5.55)$$

$$v_m = \frac{v_{i,j+1}^{n} + v_{i,j}^{n} + v_{i-1,j+1}^{n} + v_{i-1,j}^{n}}{4} \qquad (5.56)$$

$$h_{mx} = \frac{h_{i,j}^{n} + h_{i-1,j}^{n}}{2} \qquad (5.57)$$

$$h_{my} = \frac{h_{i,j}^{n} + h_{i,j-1}^{n}}{2} \qquad (5.58)$$

In the case of a wave-generated circulation, the components of the radiation stresses (S_{ij}) can be computed in the centre of the grid cells, using the mean over-the-cell 'i-j' values: $\frac{u_{i,j}^{n} + u_{i+1,j}^{n}}{2}$, $\frac{v_{i,j}^{n} + v_{i,j+1}^{n}}{2}$, and $\zeta_{i,j}^{n}$.

Example 5.4

This exercise investigates the flow pattern in a coastal region of constant depth, where a breakwater for a marina project has been built. The solution domain is 350 m × 200 m discretized into a 70 × 40 rectangular grid. The incoming from the left longshore current has a constant velocity of 0.2 m/s. The right boundary is described by a radiation (free-outflow) condition. The same free-outflow condition is applied to the open-sea boundary condition. The input data used for the simulation are as follows:

Water depth = 4 m
Smagorinsky parameter = 0.3
Coriolis coefficient = 0.0001 s^{-1}
Bed friction = 0.001

The shape of the structure is provided in the text file (depths.txt). The grid cells occupied by the coastline and the breakwater structure are defined as zero depth, and the rest of the cells as 4 (indicative of the water depth). The spatial discretization steps were $\Delta x = \Delta y = 5$ m and the time step $\Delta t = 0.5$ s.

The computations started with a cold start where velocities and water surface fluctuations were taken as zeros. The simulation was conducted for 14,400 time steps (2.0 hours), since before that time the system had not reached a steady-state condition. This is evidenced by the time evolution of the total kinetic energy where near the end of the simulation it started stabilizing (Figure 5.13). The flow field is illustrated in Figure 5.14. The circulation patterns in terms of velocity magnitude and vortex patterns can facilitate studies related to contaminant dispersion, sediment transport, foundation scouring and vessel navigability.

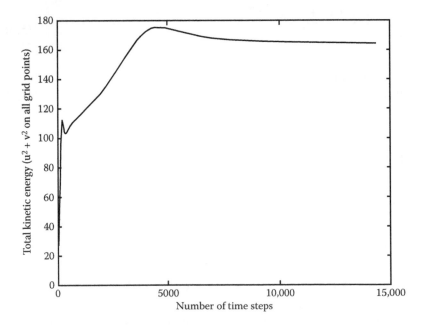

Figure 5.13 Evolution of the total kinetic energy in time.

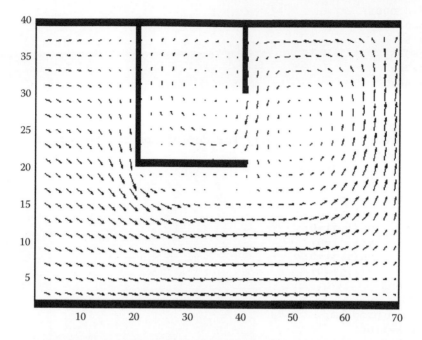

Figure 5.14 Circulation pattern around the marina breakwater.

Computer code 5.4

```
% Example 5.4 Longshore Current Past a Marina Breakwater
% ho = Water depth [m];
% u = Velocity in the x-direction [m/s];
% v = Velocity in the y-direction [m/s];
% cv = Longshore current velocity [m/s];
% ev = Eddy viscosity [m^2/s];
% sp = Smagorinsky parameter for horizontal eddy coefficient
[dimensionless];
% cf = Coriolis coefficient [1/s];
% fb = bed friction [dimensionless];
% Dd = Spatial step in both x and y directions [m];
% Dt = Time step [s];
% nx = Number of computational steps in the x-direction;
% ny = Number of computational steps in the y-direction;
% nt = Number of time steps;
clc; clear all; close all;
% Input data;
g=9.81;
ho=4;
cv=0.2;
sp=0.3;
```

```
cf=0.0001;
fb=0.001;
Dd=5;
Dt=0.5;
nx=70;
ny=40;
nt=14400;
load depths.txt
df=depths;
% Subroutine h=fm(i,j,df): For definition of the marina
breakwater;
% Initial conditions;
for i=1:nx
    for j=1:ny
        u(i,j)=0;
        un(i,j)=0;
        v(i,j)=0;
        vn(i,j)=0;
        z(i,j)=0;
    end
end
% Main program;
for k=1:nt
    % Estimation of the eddy viscosity;
    for i=2:nx-1
        for j=2:ny-1
            t1=(u(i+1,j)-u(i,j))/Dd-(v(i,j+1)-v(i,j))/Dd;
            t2=(u(i,j+1)+u(i+1,j+1)-u(i,j-1)-u(i+1,j-1))/2/
            Dd+(v(i+1,j+1)+v(i+1,j)-v(i-1,j+1)-v(i-1,j))/4/Dd;
            ev(i,j)=sp*Dd^2*sqrt(t1^2+t2^2);
            if ev(i,j) < 0.05;
                ev(i,j)=0.05;
            end
        end
    end
    % Solution of the continuity equation;
    for i=2:nx-2
        for j=2:ny-2
            h1=fm(i,j,df);
            h2=fm(i-1,j,df);
            h3=fm(i+1,j,df);
            h4=fm(i,j+1,df);
            h5=fm(i,j-1,df);
            if h1>0.1
                hl=(h1+h2)/2;
                hr=(h1+h3)/2;
                ho=(h4+h1)/2;
                hu=(h1+h5)/2;
                z(i,j)=z(i,j)-Dt/Dd*(u(i+1,j)*hr-
                u(i,j)*hl+v(i,j+1)*ho-v(i,j)*hu);
```

```
                end
            end
        end
        % Solution x-axis equilibrium equation;
        for i=3:nx-2
            for j=2:ny-2
                evd=(ev(i,j)+ev(i-1,j))/2;
                h1=fm(i,j,df);
                h2=fm(i-1,j,df);
                if h1>0 && h2>0
                    hm=(h1+h2)/2;
                    vm=(v(i,j)+v(i,j+1)+v(i-1,j)+v(i-1,j+1))/4;
                    adv=-u(i,j)*(u(i+1,j)-u(i-1,j))/2/
                    Dd-vm*(u(i,j+1)-u(i,j-1))/2/Dd;
                    cor=vm*cf;
                    frict=-fb*sqrt(u(i,j)^2+vm^2)*u(i,j)/hm;
                    pres=-g*(z(i,j)-z(i-1,j))/Dd;
                    diff=evd*(-4*u(i,j)+u(i+1,j)+u(i-
                    1,j)+u(i,j+1)+u(i,j-1))/Dd^2;
                    un(i,j)=u(i,j)+Dt*(cor+pres+frict+diff+adv);
                end
            end
        end
        % Boundary conditions for velocity u;
        for j=2:ny-2
            h8=fm(2,j,df);
            if h8>0.1
                % Constant flow boundary condition;
                un(2,j)=cv;
            end
            h9=fm(nx-2,j,df);
            if h9>0.1
                % Open sea boundary condition;
                un(nx-1,j)=z(nx-2,j)*sqrt(g/h9);
            end
        end
        for i=2:nx-2
            un(i,1)=un(i,2);
            un(i,ny-1)=un(i,ny-2);
        end
        % Solution of the y-axis equilibrium condition;
        for i=2:nx-2
            for j=3:ny-1
                edd=(ev(i,j)+ev(i,j-1))/2;
                h1=fm(i,j,df);
                h5=fm(i,j-1,df);
                if h1>0 && h5>0
                    hm=(h1+h5)/2;
                    um=(u(i,j)+u(i+1,j)+u(i,j-1)+u(i+1,j-1))/4;
```

```
                        adv=-v(i,j)*(v(i,j+1)-v(i,j-1))/2/
                        Dd-um*(v(i+1,j)-v(i-1,j))/2/Dd;
                        cor=-um*cf;
                        frict=-fb*sqrt(v(i,j)^2+um^2)*v(i,j)/hm;
                        pres=-g*(z(i,j)-z(i,j-1))/Dd;
                        diff=evd*(-4*v(i,j)+v(i,j-1)+v(i,j+1)+v(i-
                        1,j)+v(i+1,j))/Dd^2;
                        vn(i,j)=v(i,j)+Dt*(cor+pres+frict+diff+adv);
                end
        end
    end
    % Open sea boundary conditions for velocity v:
    for i=2:nx-2
        h6=fm(i,2,df);
        h7=fm(i,ny-2,df);
        if h6<0.1
            if h7<0.1
                continue;
            else
                vn(i,ny-1)=z(i,ny-2)*sqrt(g/h7);
            end
        end
        vn(i,2)=-z(i,2)*sqrt(g/h6);
    end
    for j=2:ny-2
        vn(1,j)=vn(2,j);
        vn(nx-1,j)=vn(nx-2,j);
    end
    % Updating the velocities u and v;
    for i=1:nx
        for j=1:ny
            u(i,j)=un(i,j);
            v(i,j)=vn(i,j);
            % Zero flux boundary condition;
        end
    end
    % Calculation of the kinetic energy;
    ke=0;
    for i=1:nx
        for j=1:ny
            ke=ke+u(i,j)^2+v(i,j)^2;
        end
    end
    KinE(k)=ke;
    %        index=k
end
plot(1:k,KinE,'Color','k','Linewidth',1.5)
xlabel('Number of time steps')
ylabel('Total kinetic energy (u^2+v^2 on all grid points)')
```

```
[i,j]=meshgrid(1:1:nx,1:1:ny);
df1=df(:,3);
df2=reshape(df1,70,40);
df2=df2';
An=ones(size(df2));
idx=find(df2==0);
An(idx)=0;
figure,pcolor(An); hold on;
colormap(bone)
shading flat
% Plot every interval;
% quiver(i,j,u',v','Color','k');
% Plot every second interval;
quiver(i(1:2:end,1:2:end),j(1:2:end,1:2:end),(u(1:2:end,1:2:
end))',(v(1:2:end,1:2:end))','Color','k'); hold on;
```

PROBLEM 5.4
Solve the same problem by making the following suggested changes while keeping the rest of the data constant:

1. Change the incoming longshore velocity (left-hand boundary), so that it linearly varies from 0 at the coastline to 0.4 m/s at the open-sea boundary. Run the simulation and discuss the results as compared to those of the constant longshore velocity.
2. Change linearly the water depth from 4 metres at the coastline to 8 metres at the open-sea boundary. Run the simulation and discuss the results as compared to those of constant depth domain.
3. Remove the entrance jetty and conduct the simulation with only the L-shaped breakwater. Compare the simulation results with those from the original breakwater configuration.
4. Change the Smagorinsky parameter to 0.1 and 1.0. Run the simulations, compare and discuss the results.
5. Shorten the horizontal arm of the breakwater from 100 m to 50 m, while keeping intact the rest of the configuration. Run the program and compare the simulation results with those from the original breakwater configuration.

Example 5.5

This exercise investigates the wind-generated flow pattern in the Thermaikos Gulf, near the City of Thessaloniki, Greece (Figure 5.15). The topography is given in the text file ThermD.txt. The data are provided for a rectangular domain of 44 km × 46 km, on a square grid of 2 km × 2 km. The inland 'dry' areas are defined by a zero elevation, and the gulf water depths by a positive number for every cell center (i,j). The radiation condition is applied to the open-sea boundary, and

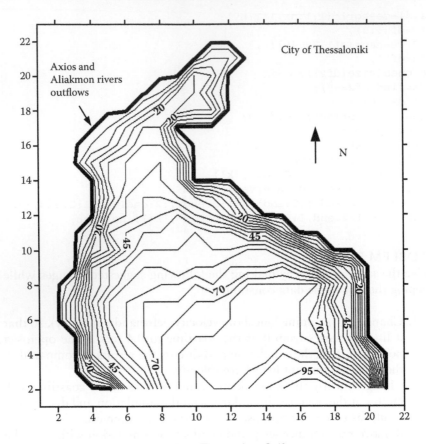

Figure 5.15 Bathymetry contours of the Thermaikos Gulf.

a zero-normal-flux condition is applied to the coastal boundaries. The input data used for the simulation are as follows:

Wind velocity in the x-direction = 7 m/s
Wind velocity in the y-direction = −7 m/s
Smagorinsky parameter = 0.3
Coriolis coefficient = 0.0001 s^{-1}
Bed friction coefficient = 10^{-4}
Water surface friction coefficient = 3 × 10^{-6}
Low limit for the horizontal eddy coefficient = 1.0 m^2/s
The spatial discretization steps were $\Delta x = \Delta y$ = 2000 m and the time step Δt = 30 s

The simulation was initiated with cold-start conditions and was conducted for 21,600 time steps (7.5 days). A three-dimensional rendering of the gulf topography is provided in Figure 5.16.

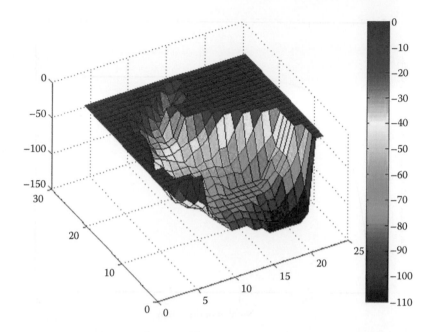

Figure 5.16 Three-dimensional illustration of the Thermaikos Gulf topography.

The simulation period was sufficient for the establishment of steady-state conditions. As can be seen from the time evolution of the total kinetic energy graph (Figure 5.17), steady-state conditions were reached after about 11,520 time steps (4 days).

The effects of the continuously blowing northwestern wind result into the creation of two distinct large-scale clockwise vortices generating a northbound current drift along the west coast and a southbound drift along the east coast of the gulf (Figure 5.18).

The three-dimensional nature of the flow in such sea basins is sometimes important, and particularly in the case of wind-generated currents. In order to avoid the application of 3-D models, one can adopt a functional form of the velocity variation along the depth (a typical velocity profile) and deduce the local velocities $U(x,y,z,t)$, $V(x,y,z,t)$ from their depth mean values $u(x,y,t)$, $v(x,y,t)$, the water depth $h(x,y)$ and the wind friction τ_{sx}, τ_{sy} (functions of w_x, w_y). If a parabolic velocity distribution is assumed, then the following relations can be applied for the velocity U (Koutitas and Gousidou-Koutita 1986):

$$U = \alpha_1 z^2 + \alpha_2 z + \alpha_3 \quad \text{where} \quad -h \leq z \leq 0 \tag{5.59}$$

where the coefficients α_1, α_2 and α_3 are defined as

$$\alpha_1 = \frac{3}{4h}\frac{\tau_{sx}}{\rho\varepsilon_h} - \frac{3}{2}\frac{u}{h^2} \tag{5.60}$$

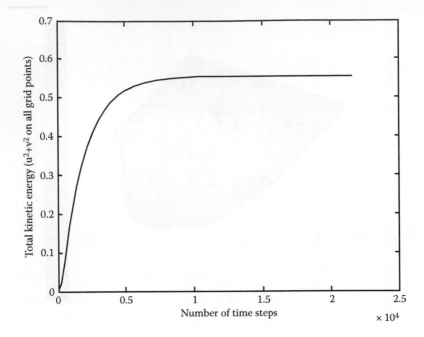

Figure 5.17 Evolution of the total kinetic energy in time.

$$\alpha_2 = \frac{\tau_{sx}}{\rho\varepsilon_h} \tag{5.61}$$

$$\alpha_3 = \frac{h}{4\varepsilon_h}\frac{\tau_{sx}}{\rho} + \frac{3}{2}u \tag{5.62}$$

Similarly the equations for V are

$$V = \beta_1 z^2 + \beta_2 z + \beta_3 \text{ where } -h \leq z \leq 0 \tag{5.63}$$

$$\beta_1 = \frac{3}{4h}\frac{\tau_{sy}}{\rho\varepsilon_h} - \frac{3}{2}\frac{v}{h^2} \tag{5.64}$$

$$\beta_2 = \frac{\tau_{sy}}{\rho\varepsilon_h} \tag{5.65}$$

$$\beta_3 = \frac{h}{4\varepsilon_h}\frac{\tau_{sy}}{\rho} + \frac{3}{2}v \tag{5.66}$$

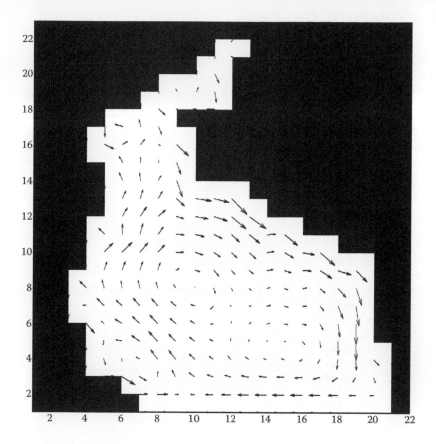

Figure 5.18 Wind-induced current circulation pattern.

Computer code 5.5

```
% Example 5.5 Wind Circulation in Thermaikos Gulf
% ho = Water depth [m];
% u = Velocity in the x-direction [m/s];
% v = Velocity in the y-direction [m/s];
% wx = Wind velocity along the x-axis [m/s];
% wy = Wind velocity along the y-axis [m/s];
% dif = Lower limit of the eddy viscosity;
% cf = Coriolis coefficient [1/s];
% fs = Water surface friction coefficient;
% fb = Bed friction coefficient;
% sp = Smagorinsky parameter for horizontal eddy coefficent
[dimensionless];
% Dd = Spatial step in both x and y directions [m];
% Dt = Time step [s];
% nx = Number of computational steps in the x-direction;
% ny = Number of computational steps in the y-direction;
```

```
% nt = Number of time steps;
clc; clear all; close all;
% Input data;
g=9.81;
ho=0;
wx=7;
wy=-7;
dif=1;
cf=0.0001;
fs=0.000003;
fb=0.01;
sp=0.3;
Dd=2000;
Dt=30;
nx=22;
ny=23;
nt=21600;
load ThermD.txt
df=ThermD;
% Import h=fm(i,j,df): Subroutine that defines depths:
% Initial conditions;
for i=1:nx
    for j=1:ny
        u(i,j)=0;
        un(i,j)=0;
        v(i,j)=0;
        vn(i,j)=0;
        z(i,j)=0;
    end
end
% Main program;
for k=1:nt
    % Estimation of the eddy viscosity;
    for i=2:nx-1
        for j=2:ny-1
            t1=(u(i+1,j)-u(i,j))/Dd-(v(i,j+1)-v(i,j))/Dd;
            t2=(u(i,j+1)+u(i+1,j+1)-u(i,j-1)-u(i+1,j-1))/2/Dd+
            (v(i+1,j+1)+v(i+1,j)-v(i-1,j+1)-v(i-1,j))/4/Dd;
            ev(i,j)=sp*Dd^2*sqrt(t1^2+t2^2);
            if ev(i,j) < 1;
                ev(i,j)=1;
            end
        end
    end
    % Solution of the continuity equation;
    for i=2:nx-2
        for j=2:ny-2
            h1=fm(i,j,df);
            h2=fm(i-1,j,df);
            h3=fm(i+1,j,df);
```

```
            h4=fm(i,j+1,df);
            h5=fm(i,j-1,df);
            if h1>0.1
                hl=(h1+h2)/2;
                hr=(h1+h3)/2;
                ho=(h4+h1)/2;
                hu=(h1+h5)/2;
                z(i,j)=z(i,j)-Dt/Dd*(u(i+1,j)
                *hr-u(i,j)*hl+v(i,j+1)*ho-v(i,j)*hu);
            end
        end
    end
% Solution x-axis equilibrium equation;
for i=3:nx-2
    for j=2:ny-2
        evd=(ev(i,j)+ev(i-1,j))/2;
        h1=fm(i,j,df);
        h2=fm(i-1,j,df);
        if h1>0 && h2>0
            hm=(h1+h2)/2;
            vm=(v(i,j)+v(i,j+1)+v(i-1,j)+v(i-1,j+1))/4;
            surfx=fs*wx*sqrt(wx^2+wy^2)/hm;
            adv=-u(i,j)*(u(i+1,j)-u(i-1,j))/2/
            Dd-vm*(u(i,j+1)-u(i,j-1))/2/Dd;
            cor=vm*cf;
            frict=-fb*sqrt(u(i,j)^2+vm^2)*u(i,j)/hm;
            pres=-g*(z(i,j)-z(i-1,j))/Dd;
            diff=evd*(-4*u(i,j)+u(i+1,j)+u(i-
            1,j)+u(i,j+1)+u(i,j-1))/Dd^2;
            un(i,j)=u(i,j)+Dt*(surfx+cor+pres+frict+diff
            +adv);
        end
    end
end
% Boundary conditions for velocity u;
for j=2:ny-2
    h8=fm(2,j,df);
    if h8>0.1
        un(2,j)=-z(2,j)*sqrt(g/h8);
    end
    h9=fm(nx-2,j,df);
    if h9>0.1
        un(nx-1,j)=z(nx-2,j)*sqrt(g/h9);
    end
end
for i=2:nx-2
    un(i,1)=un(i,2);
    un(i,ny-1)=un(i,ny-2);
end
% Solution of the y-axis equilibrium condition;
```

```
for i=2:nx-2
    for j=3:ny-1
        edd=(ev(i,j)+ev(i,j-1))/2;
        h1=fm(i,j,df);
        h5=fm(i,j-1,df);
        if h1>0 && h5>0
            hm=(h1+h5)/2;
            um=(u(i,j)+u(i+1,j)+u(i,j-1)+u(i+1,j-1))/4;
            surfy=fs*wy*sqrt(wx^2+wy^2)/hm;
            adv=-v(i,j)*(v(i,j+1)-v(i,j-1))/2/
            Dd-um*(v(i+1,j)-v(i-1,j))/2/Dd;
            cor=-um*cf;
            frict=-fb*sqrt(v(i,j)^2+um^2)*v(i,j)/hm;
            pres=-g*(z(i,j)-z(i,j-1))/Dd;
            diff=evd*(-4*v(i,j)+v(i,j-1)+v(i,j+1)+v(i-
            1,j)+v(i+1,j))/Dd^2;
            vn(i,j)=v(i,j)+Dt*(surfy+cor+pres+frict+diff
            +adv);
        end
    end
end
% Boundary conditions for velocity v:
for i=2:nx-2
    h6=fm(i,2,df);
    if h6>0.1
        vn(i,2)=-z(i,2)*sqrt(g/h6);
    end
    h7=fm(i,ny-2,df);
    if h7>0.1
        vn(i,ny-1)=z(i,ny-2)*sqrt(g/h7);
    end
end
for j=2:ny-2
    vn(1,j)=vn(2,j);
    vn(nx-1,j)=vn(nx-2,j);
end
% Updating the velocities u and v;
for i=1:nx
    for j=1:ny
        u(i,j)=un(i,j);
        v(i,j)=vn(i,j);
    end
end
% Calculation of the kinetic energy;
ke=0;
for i=1:nx
    for j=1:ny
        ke=ke+u(i,j)^2+v(i,j)^2;
    end
end
```

```
      KinE(k)=ke;
      index=k
end
[X,Y]=meshgrid(1:1:nx,1:1:ny);
[m n]=size(Y);
for a=1:m
      for b=1:n
            Z(a,b)=-fm(b,a,df);
      end
end
surf(X,Y,Z)
figure
plot(1:k,KinE,'Color','k','Linewidth',1.5);
xlabel('Number of time steps')
ylabel('Total kinetic energy (u^2+v^2 on all grid points)')
[i,j]=meshgrid(1:1:nx,1:1:ny);
df1=df(:,3);
df2=reshape(df1,22,23);
df2=df2';
An=ones(size(df2));
idx=find(df2==0);
An(idx)=0;
figure,pcolor(An); hold on;
colormap(bone)
shading flat
LL=quiver(i,j,u',v','Color','k');
% set(LL,'linewidth',1.5);
```

PROBLEM 5.5

Solve the same application of the Thermaikos Gulf by making the following suggested changes while keeping the rest of the data constant:

1. Change the wind data to $w_x = -10$ m/s and $w_y = 10$ m/s (southeastern wind) while increasing the wind surface friction to $f_s = 10^{-5}$. Run the simulation and comment on the data.
2. Set the Smagorinsky parameter equal to 0.1 and 1.0. Run the simulation for the two different values and comment on the results obtained.
3. Change the time step to $\Delta t = 60$ s and $\Delta t = 600$ s. Run the simulations, and then compare and discuss the results.
4. Modify the original code and data so that the model simulates tidal fluctuations introduced at the open-sea boundary as $u(x, y = 0, t) = 0$ and $v(x, y = 0, t) = a_o \sin\left(\dfrac{2\pi t}{T}\right)$, where $a_o = 0.5$ m and $T = 12.0$ hr. Run the simulation, then evaluate and discuss the results.
5. Modify the original code by including Equations 5.59 to 5.66 to determine the vertical velocity distribution at the nodal point ($i = 13$, $j = 8$) (see Figure 5.18). Comment on the distribution.

5.4 STRATIFIED FLOWS IN GEOPHYSICAL DOMAINS

A very important class of free surface flows in geophysical flow domains such as coastal and ocean waters are characterized by thermal, saline or fluid-mud stratification. The first results mainly from the heating of surface water masses during the hot period of the year; the second results from the fresh water outflows from rivers and creeks to the coast; the third results from underwater mudslides or even re-suspension of deposited sediments.

Although stratification is a complicated three-dimensional phenomenon, it is common to assume that the two layers of different density are separated by a sharp interface. That interface is known as thermocline for thermal stratification, halocline for saline stratification and lutocline for mud-induced stratification. This thin-interface simplification negates the interfacial mixing. This is a valid assumption for time scales smaller than the time required for considerable mixing of the two layers. This mixing is locally reduced, due to the turbulence suppression in the interfacial region, as described by the Richardson number. If the density of the upper layer is ρ_o and the density of the lower layer is ρ_u, the system is hydrodynamically stable if $\rho_o < \rho_u$ (Scarlatos 1996b).

The distinction of the two layers permits the assumption of nearly horizontal flow in each layer, and the description of the layer hydrodynamics via hydrostatic pressure distribution and layer-mean flow velocities u_o and u_u, respectively.

The frictional forces that act on and mobilize the two layers are

- The free surface shear exercised by the wind, τ_s, related to the wind velocity w (usually measured 10 m above sea level) as

$$\frac{\tau_s}{\rho_o} = f_s w |w| \tag{5.67}$$

- The bottom shear due to the solid bed boundary, related to the lower layer velocity as

$$\frac{\tau_b}{\rho_u} = f_b u_u |u_u| \tag{5.68}$$

- The interfacial shear due to the velocity differences $u_o - u_u$ expressed as

$$\frac{\tau_i}{\rho_o} = f_i (u_o - u_u) |u_o - u_u| \tag{5.69}$$

where f_i is the interfacial friction coefficient.

The friction coefficients f_s, f_i and f_b depend collectively on the wind, flow intensity, fluid properties and the bed roughness. Average values are of the order $O(f_s) = 3 \times 10^{-6}$, $O(f_i) = 10^{-3}$ and $O(f_b) = 10^{-2}$.

For a two-dimensional x–y horizontal domain, the unknown variables are the layer total depths h_o and h_u, and the layer mean velocities u_o, u_u, v_o and v_u, all functions of x, y and t. The equations formulated for the estimation of the aforementioned functions are based on the principles of mass continuity and the forces equilibrium along the horizontal directions (Savvidis et al. 2004).

5.4.1 Governing equations for horizontal two-dimensional stratified flows

By assuming small water surface and interface slopes, the equations for the upper and lower layers can be derived similarly to the ones for quasi-horizontal two-dimensional geophysical flows. Thus, by following the same procedure as for the depth-averaged free surface flows, the governing equations for the two layers in the two horizontal directions are

Upper layer

$$\frac{\partial h_o}{\partial t} + \frac{\partial (h_o u_o)}{\partial x} + \frac{\partial (h_o v_o)}{\partial y} = 0 \tag{5.70}$$

$$\frac{\partial u_o}{\partial t} + u_o \frac{\partial u_o}{\partial x} + v_o \frac{\partial u_o}{\partial y} = -g \frac{\partial (h_o + h_u + z_b)}{\partial x} + f v_o$$
$$+ \varepsilon_h \left(\frac{\partial^2 u_o}{\partial x^2} + \frac{\partial^2 u_o}{\partial y^2} \right) + \frac{\tau_{sx}}{\rho_o h_o} - \frac{\tau_{ix}}{\rho_o h_o} \tag{5.71}$$

$$\frac{\partial v_o}{\partial t} + u_o \frac{\partial v_o}{\partial x} + v_o \frac{\partial v_o}{\partial y} = -g \frac{\partial (h_o + h_u + z_b)}{\partial y} - f u_o$$
$$+ \varepsilon_h \left(\frac{\partial^2 v_o}{\partial x^2} + \frac{\partial^2 v_o}{\partial y^2} \right) + \frac{\tau_{sy}}{\rho_o h_o} - \frac{\tau_{iy}}{\rho_o h_o} \tag{5.72}$$

Lower layer

$$\frac{\partial h_u}{\partial t} + \frac{\partial (h_u u_u)}{\partial x} + \frac{\partial (h_u v_u)}{\partial y} = 0 \tag{5.73}$$

$$u_u \frac{\partial u_u}{\partial x} + v_u \frac{\partial u_u}{\partial y} = -g \frac{\partial[(h_o + h_u + z_b) - \Delta\rho h_o]}{\partial x} + fv_u$$

$$+ \varepsilon_h \left(\frac{\partial^2 u_u}{\partial x^2} + \frac{\partial^2 u_u}{\partial y^2} \right) + \frac{\tau_{ix}}{\rho_o h_o} - \frac{\tau_{bx}}{\rho_u h_u}$$

(5.74)

$$\frac{\partial v_u}{\partial t} + u_u \frac{\partial v_u}{\partial x} + v_u \frac{\partial v_u}{\partial y} = -g \frac{\partial\left[(h_o + h_u + z_b) - \Delta\rho h_o\right]}{\partial y} - fu_u$$

$$+ \varepsilon_h \left(\frac{\partial^2 v_u}{\partial x^2} + \frac{\partial^2 v_u}{\partial y^2} \right) + \frac{\tau_{iy}}{\rho_o h_o} - \frac{\tau_{by}}{\rho_u h_u}$$

(5.75)

where $\Delta\rho$ is the relative density difference

$$\Delta\rho = \frac{\rho_u - \rho_o}{\rho_u}$$

(5.76)

The numerical solution is accomplished similarly to the previous nearly horizontal free surface flow models, by using a FTCS (forward in time, central in space) explicit scheme on a staggered computational grid, through successive computations of the velocity and depth variables.

5.4.2 One-dimensional stratified system

For the case of a one-dimensional stratified system in a narrow basin confined by vertical shore boundaries (full reflection) the governing equations (Equations 5.70 to 5.76) are reduced as follows:

$$\frac{\partial h_o}{\partial t} + \frac{\partial(h_o u_o)}{\partial x} = 0$$

(5.77)

$$\frac{\partial u_o}{\partial t} + u_o \frac{\partial u_o}{\partial x} = -g \frac{\partial(h_o + h_u + z_b)}{\partial x} + \varepsilon_h \frac{\partial^2 u_o}{\partial x^2} + \frac{\tau_{sx}}{\rho_o h_o} - \frac{\tau_{ix}}{\rho_o h_o}$$

(5.78)

$$\frac{\partial h_u}{\partial t} + \frac{\partial(h_u u_u)}{\partial x} = 0$$

(5.79)

$$\frac{\partial u_u}{\partial t} + u_u \frac{\partial u_u}{\partial x} = -g \frac{\partial \left[(h_o + h_u + z_b) - \Delta \rho h_o \right]}{\partial x}$$

$$+ \varepsilon_h \frac{\partial^2 u_u}{\partial x^2} + \frac{\tau_{ix}}{\rho_o h_o} - \frac{\tau_{bx}}{\rho_u h_u}$$

(5.80)

When subject to wind-induced stress on the open surface, the system is mobilized. The unknown quantities describing the phenomenon are the water velocities u_o and u_u, and total depths h_o and h_u (Figure 5.19).

During the initial mobilization of the two layers, a surface wave and an interfacial wave are generated and oscillate until steady-state conditions are established. The phenomenon of upwelling, that is, suppression of the lower layer on one end of the basin and appearance of that lower layer on the surface at the opposite end, is not described by the model, as situations of $h_o = 0$ or $h_u = 0$ are not provided.

Example 5.6

This exercise investigates the wind-generated flow pattern in a one-dimensional stratified two-layer system. The system is initially at rest and then is subject to continuous wind stress applied at the free water surface. The data used for this exercise are as follows:

Wind velocity = 10 m/s
Open surface wind friction coefficient = 3×10^{-6}
Interface friction coefficient = 0.001
Bed friction coefficient = 0.05
Eddy viscosity = 20 m²/s
Relative density difference = 0.005

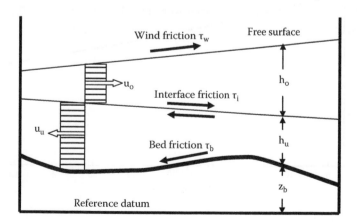

Figure 5.19 One-dimensional configuration of a two-layer stratified system.

Total length of the system = 10000 m
Equilibrium depth of the upper layer = 10 m
Equilibrium depth of the lower layer = 10 m

The computational discretization steps are Δx = 200 m and Δt = 10 s. The simulation was conducted for a total of 103,680 time steps (12 days). The tilting of the stabilized interface in the direction of the wind is shown in Figure 5.20. The fact that the interface has been stabilized is documented by the history of the water depth changes of the two fluid layers at the far end of the basin (Figure 5.21).

Computer code 5.6

```
% Example 5.6 Wind Induced Upwelling in Stratified Water
% w = Wind velocity [m/s];
% fs = Friction coefficient at water surface;
% fi = Friction coefficient at fluid interface;
% fb = Friction coefficient at bed;
% rdd = Relative density difference of the fluids;
% ev = Eddy viscosity coefficient [m^2/s];
% huo = Initial upper fluid layer depth [m];
```

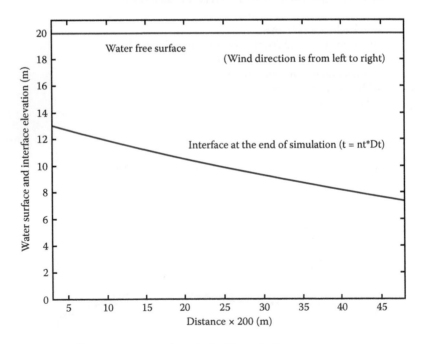

Figure 5.20 Stabilized fluid interface after a 12-day time period.

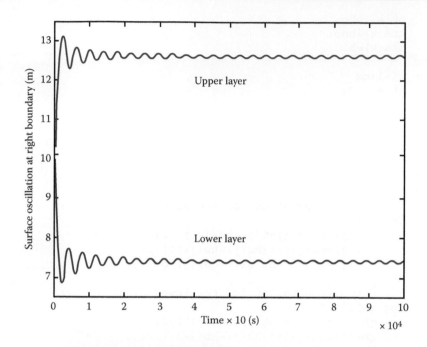

Figure 5.21 Oscillations of the free surface and interface at the end boundary.

```
% hbo = Initial lower fluid layer depth [m];
% Dx = Spatial step [m];
% Dt = Time step [s];
% nx = Number of spatial computational steps;
% nt = Number of time computational steps;
clc; clear all; close all;
% Input data;
g=9.81;
w=10;
fs=0.000003;
fi=0.001;
fb=0.05;
rdd=0.005;
ev=20;
huo=10;
hbo=10;
Dx=200;
Dt=10;
nx=50;
nt=103680;
% Initial conditions;
for k=1:nx-1
    hu(k)=huo;
```

```
    hb(k)=hbo;
    hun(k)=huo;
    hbn(k)=hbo;
end
for k1=1:nx
    zb(k1)=0;
    uu(k1)=0;
    ub(k1)=0;
    uun(k1)=0;
    ubn(k1)=0;
end
% Main program;
for i=1:nt
    % Continuity equation for the upper fluid layer;
    for j=2:nx-2
        qul(j)=uu(j)*(hu(j)+hu(j-1))/2;
        qur(j)=uu(j+1)*(hu(j)+hu(j+1))/2;
        hun(j)=hu(j)-(Dt/Dx)*(qur(j)-qul(j));
    end
    % Continuity equation for the lower fluid layer;
    for j=2:nx-2
        qbl(j)=ub(j)*(hb(j)+hb(j-1))/2;
        qbr(j)=ub(j+1)*(hb(j)+hb(j+1))/2;
        hbn(j)=hb(j)-(Dt/Dx)*(qbr(j)-qbl(j));
    end
    % Updating of layer depth values;
    for j=2:nx-2
        hu(j)=hun(j);
        hb(j)=hbn(j);
    end
    % Force equilibrium equation for the upper layer;
    for j=3:nx-2
        hum(j)=(hu(j)+hu(j-1))/2;
        ts(j)=fs*w*abs(w)/hum(j);
        ti(j)=fi*(uu(j)-ub(j))*abs(uu(j)-ub(j))/hum(j);
        ed(j)=ev*(uu(j-1)-2*uu(j)+uu(j+1))/Dx^2;
        uun(j)=uu(j)-Dt*g*(hu(j)+hb(j)+zb(j)-hu(j-1)-hb(j-
        1)-zb(j-1))/Dx+Dt*(ed(j)+ts(j)-ti(j));
    end
    % Force equilibrium equation for the lower layer;
    for j=3:nx-2
        hbm(j)=(hb(j)+hb(j-1))/2;
        ti(j)=fi*(uu(j)-ub(j))*abs(uu(j)-ub(j))/hbm(j);
        tb(j)=fb*ub(j)*abs(ub(j))/hbm(j);
        ed(j)=ev*(ub(j-1)-2*ub(j)+ub(j+1))/Dx^2;
        ubn(j)=ub(j)-Dt*g*((hu(j)+hb(j)+zb(j)-hu(j-1)-hb(j-
        1)-zb(j-1))/Dx-rdd*(hu(j)-hu(j-1))/
        Dx)+Dt*(ed(j)+ti(j)-tb(j));
    end
    % Updating of fluid velocity values;
```

```
    for j=3:nx-2
        uu(j)=uun(j);
        ub(j)=ubn(j);
        Hbfinal(j)=hbn(j);
        Htotal(j)=hun(j)+hbn(j);
    end
    Eu=0;
    El=0;
    for j=3:nx-2
        Eu=Eu+(hu(j)+hu(j-1))/2*uu(j)^2;
        El=El+(hb(j)+hb(j-1))/2*ub(j)^2;
    end
    Hupper(i)=hu(nx-2);
    Hlower(i)=hb(nx-2);
end
plot(1:j,Hbfinal,'r','Linewidth',1.5)
hold on
plot(1:j,Htotal,'b','Linewidth',1.5)
axis([3,48,0,21])
xlabel('Distance x 200 [m]'), ylabel('Water surface and
interface elevation [m]')
text(10,18.9,'Water free-surface')
text(20,11.5,'Interface at the end of simulation (t =
nt*Dt)')
text(25,18,'(Wind direction is from left to right)')
figure, plot(1:i,Hupper,'b','Linewidth',1.5)
hold on
plot(1:i,Hlower,'r','Linewidth',1.5)
axis([0,100000,6.5,13.5])
xlabel('Time x 10 [s]'), ylabel('Surface oscillation at
right boundary [m]')
text(40000,12,'Upper layer')
text(40000,8,'Lower layer')
```

PROBLEM 5.6

Solve the same exercise by making the following suggested changes while keeping the rest of the data constant:

1. Change the relative density difference to 0.01, 0.001 and 0.0001. Run the simulations until steady-state conditions are achieved. Compare and comment on your results. Note that full upwelling may destabilize the solution.
2. Change the wind speed to 1 m/s, 5 m/s, 25 m/s and 50 m/s. Run the simulations until steady-state conditions are achieved. Compare and comment on your results.
3. Change the code as needed, so that the program simulates a left-to-right wind of 10 m/s for 6 hours and then a right-to-left wind of

the same magnitude for another 6 hours. Run the simulation for a total of 12 hours and comment on the results obtained at 1-hour time intervals.

4. Change the equilibrium depths first to $h_o = 5$ m and $h_u = 15$ m, and then to $h_o = 15$ m and $h_u = 5$ m. Conduct the simulations and compare the results of those two scenarios.

5. Change the eddy viscosity to 1, 10 and 30. Run the simulations until steady-state conditions are achieved. Compare and comment on your results.

Chapter 6

Surface gravity water waves

6.1 BASIC CONCEPTS OF WATER WAVES

This chapter covers the subject of water waves, that is, the propagation and transformation of surface water waves driven by the gravitational force. Gravity water waves constitute part of the free surface, unsteady (usually periodic) class of flows. Traditionally this subject is covered under the disciplines of coastal, ocean or maritime engineering (U.S. Army Corps of Engineers 2002; Kim 2009).

The forms of the mathematical models to be presented and numerically solved are the simplest ones, though useful and operational. Simplifying assumptions such as linearity, periodicity and single wave frequency will be made in order to avoid mathematical and numerical complications. Those assumptions, however, do not affect the intended scope and pedagogical value of the book. One-dimensional models of wave propagation will be presented and applied within their validity limits. The wave models will be used to describe, even approximately, an array of interesting physical phenomena such as wave shoaling, wave breaking, partial wave reflection, absorption of wave energy due to friction and wave generation and propagation due to bed deformation (the phenomenon known as tsunami waves).

The models will be subsequently extended to two spatial dimensions in order to simulate wave modulation phenomena such as wave refraction, wave diffraction, wave breaking and so on. Understanding of those phenomena is indispensable for the design of coastal, harbour, and maritime structures (Komen et al. 1994; Lin 2008).

In the following paragraphs, some basic definitions and notions of water waves theory will be reviewed and summarized (Sorensen 2006). Wavelength, L, is the distance from crest to crest (or trough to trough), c_0 is the wave celerity (the speed of the propagation of the surface deformation, not of the water particles) and T is the wave period. These three variables are connected via the simple equation

$$L = c_0 T \tag{6.1}$$

The distance between the wave crest and the wave trough is defined as the wave height, H. In the linear theory of waves (of infinitesimal amplitude), the wave amplitude is defined as $a_o = \dfrac{H}{2}$.

The wavelength, L, is related to the water depth, h, and the wave period, T, by the relation

$$L = L_o \tanh\left(\frac{2\pi h}{L}\right) \tag{6.2}$$

where L_o is the deep-water $\left(h > \dfrac{L_o}{2}\right)$ wavelength defined as

$$L_o = \frac{gT^2}{2\pi} \tag{6.3}$$

Equation 6.2 is a transcendental equation that can only be solved by successive iterations. From the preceding equations the wave celerity can be expressed as

$$c_o = \frac{gT}{2\pi} \tanh\left(\frac{2\pi h}{L}\right) \tag{6.4}$$

Equation 6.4 implies that the waves propagate with celerity proportional to their period T $\left(\text{or their angular frequency } \sigma = \dfrac{2\pi}{T}\right)$. This is called the dispersion relation since it shows that two wave trains with periods T_1 and T_2 ($T_1 > T_2$) starting simultaneously from the same location, will disperse and separate in time, as the first wave of period T_1 will propagate with higher celerity than the second one of period T_2.

A simple, monochromatic wave (containing only one periodic component) propagating over constant water depth can be described by the equation

$$\zeta(x,t) = \frac{H}{2} \sin\left(\frac{2\pi x}{L} - \frac{2\pi t}{T}\right) = a_o \sin(kx - \sigma t) \tag{6.5}$$

where k is the wave number. If the ratio $\dfrac{H}{h}$ is small, $O(10^{-1})$, the wave is defined as a wave of infinitesimal amplitude, otherwise as a wave of finite amplitude.

The water depth is also classified as

- Deep for $h > \dfrac{L_o}{2}$ with celerity: $c_o = \dfrac{gT}{2\pi}$
- Shallow for $L > 10h$ with celerity: $c_o = \sqrt{gh}$
- Intermediate for all other cases with celerity given by Equation 6.4

The bed friction acting on a propagating wave can be described by a quadratic formula as

$$\tau_{bw} = \frac{1}{2}\rho f_w u_b^2 \tag{6.6}$$

where f_w is a friction coefficient and u_b the amplitude of the near-the-bed velocity:

$$u_b = \frac{\pi H}{T\sinh(kh)} \tag{6.7}$$

6.2 ONE-DIMENSIONAL GRAVITY LONG WAVES

By making the same assumptions as those made for the derivation of the unsteady free surface flow equations (see Chapter 5, Section 5.2.2.1) the wave equations for a unit-width, frictionless domain can be written (based on the continuity [Equation 5.26] and momentum [Equation 5.24]) as follows:

$$\frac{\partial y}{\partial t} + \frac{\partial(uy)}{\partial x} = 0 \tag{6.8}$$

$$\frac{\partial u}{\partial t} + u\frac{\partial u}{\partial x} = -g\frac{\partial y}{\partial x} + gS_o \tag{6.9}$$

By re-defining the total depth $y(x,t)$ (see Chapter 5, Figure 5.2) as

$$y(x,t) = \zeta(x,t) + h(x) \tag{6.10}$$

the system of the wave Equations 6.8 and 6.9 can be modified as

$$\frac{\partial \zeta}{\partial t} + \frac{\partial\left[(h+\zeta)u\right]}{\partial x} = 0 \tag{6.11}$$

$$\frac{\partial u}{\partial t} + \frac{1}{2}\frac{\partial u^2}{\partial x} = -g\frac{\partial \zeta}{\partial x} \qquad (6.12)$$

Due to the quadratic quantities, ζu in Equation 6.11 and u^2 in Equation 6.12, the system is known as the non-linear long-wave model. It should be noted that for long-wave (or shallow-wave, $L > 10h$) the vertical pressure is assumed to be distributed hydrostatically, since both the vertical water velocities and accelerations are small.

The non-linearity of the wave model, even after neglecting the non-linear effects of frictional energy losses, or the dispersion of the wave components in the case of a composite wave, results in the formation of a 'bore', that is, a vertical wall of water propagating with the speed of the wave celerity $c_o = \sqrt{gh}$. The phenomenon of bore formation can be easily explained by the fact that the wave celerity of a long wave depends on the water depth, while it is independent of the wave period (non-dispersive). Thus, in the case of an initially sinusoidal wave, propagating over water of constant depth, the wave crest, being on water of higher depth in comparison to the wave trough, propagates faster than the wave trough. As a result, after some time, the sinusoidal shape of the free surface is transformed to a vertical 'saw type' profile (Figure 6.1). In nature this phenomenon is mainly suppressed, due to the internal energy losses as well as to the dispersion of the wave components generated by the non-linearity of the propagating waves.

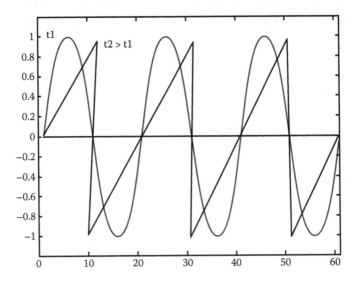

Figure 6.1 Bore formation resulting from a sinusoidal wave.

6.2.1 Linearization of the governing equations

The one-dimensional long wave model can be linearized if (1) the water depth, y, is approximated by the initial water depth, h, in the continuity equation; and (b) the non-linear advective acceleration term is neglected in the momentum equation. Then the model becomes

$$\frac{\partial \zeta}{\partial t} + \frac{\partial (uh)}{\partial x} = 0 \tag{6.13}$$

$$\frac{\partial u}{\partial t} = -g \frac{\partial \zeta}{\partial x} \tag{6.14}$$

By multiplying Equation 6.14 by h, and taking the derivate of Equation 6.14 with respect to x and of Equation 6.13 with respect to t, and subtracting the two resulting equations, a second-order hyperbolic equation is obtained, known as the telegrapher's equation (see Chapter 3, Equation 3.6). This equation can be written either for the variable $\zeta(x, t)$ (or for the $u(x, t)$) as

$$\frac{\partial^2 \zeta}{\partial t^2} - \frac{\partial}{\partial x}\left(gh \frac{\partial \zeta}{\partial x}\right) = \frac{\partial^2 \zeta}{\partial t^2} - \frac{\partial}{\partial x}\left(c_o^2 \frac{\partial \zeta}{\partial x}\right) = 0 \tag{6.15}$$

For constant depth h, Equation 6.15 collapses to Equation 3.6, where f is replaced by ζ. The telegrapher's equation describes the propagation of a wave in the positive and negative directions, with celerity depending only on the water depth (non-dispersive).

An extension of this non-dispersive wave model, in order to incorporate the influence of the wave period for intermediate and deep water, is accomplished by replacing c_o^2 with the product $c_o c_g$, where c_g is the group velocity:

$$c_g = c_o n \quad \text{where} \quad n = \frac{1}{2}\left[1 + \frac{2kh}{\sinh(2kh)}\right] \tag{6.16}$$

The Lee and Park (2001) model, formulated as an extension of the Copeland (Copeland 1958) approach, quantifies ζ in the form

$$\frac{\partial^2 \zeta}{\partial t^2} = \frac{\partial}{\partial x}\left(c_o c_g \frac{\partial \zeta}{\partial x}\right) - (\sigma^2 - k^2 c_o c_g)\zeta \tag{6.17}$$

This model can be expanded to describe the energy loss due to bed friction by adding the term $-\lambda \frac{\partial \zeta}{\partial t}$ in the right-hand side of Equation 6.17, where

λ is a dimensional coefficient (s^{-1}). Furthermore, the energy dissipation in the breaker zone due to turbulence can be modelled by introducing a term in the right-hand side of Equation 6.17 in the form of $N_b \dfrac{\partial}{\partial t}\left(\dfrac{\partial^2 \zeta}{\partial x^2}\right)$. The magnitude of the coefficient N_b is in the order of $\dfrac{1}{2}h\sqrt{gh}$ and its physical significance is analogous to the horizontal eddy viscosity coefficient.

6.2.2 Discretization of the wave equation

The numerical solution of the second-order hyperbolic equation (Equation 6.17) is usually performed using a centred finite differences scheme leading to a second-order of accuracy explicit solution (Koutitas 1988). If the values of the unknown variable $\zeta(x, t)$ at any space-time point of the computational grid ($x_i = (i - 1)\Delta t$, $t_n = (n - 1)\Delta t$) are denoted as ζ_i^n, then the discretized scheme for Equation 6.17 is written as

$$\frac{\zeta_i^{n+1} - 2\zeta_i^n + \zeta_i^{n-1}}{(\Delta t)^2} = \frac{\left(A_{i+1} + A_i\right)\left(\zeta_{i+1}^n - 2\zeta_i^n\right) - \left(A_i + A_{i-1}\right)\left(\zeta_i^n - 2\zeta_{i-1}^n\right)}{2(\Delta x)^2} - B_i\zeta_i^n$$

$$(6.18)$$

where

$$A_i = c_o c_{gi} \tag{6.19}$$

$$B_i = \sigma^2 - k_i^2 c_o c_{gi} \tag{6.20}$$

Both A_i and B_i are calculated on each grid point with depth h_i. After all of the input data are provided, Equation 6.18 can be solved explicitly in terms of the variable ζ_i^{n+1}.

An interesting feature of the solution involves the incoming wave boundary where a reflected wave ζ_{ref} is superimposed to the incident wave ζ_{inc}. The incident wave is defined by a typical sinusoidal form (see Equation 6.5), and the reflected wave, generated within the domain interior, is described by the free radiation condition known as Sommerfeld condition (see Chapter 5, Equation 5.44) that is a monochromatic wave propagating in the negative (opposing) direction with celerity c_o:

$$\frac{\partial \zeta_{rad}}{\partial t} + c_o \frac{\partial \zeta_{rad}}{\partial n} = \frac{\partial \zeta_{ref}}{\partial t} + c_o \frac{\partial \zeta_{ref}}{\partial n} = 0 \tag{6.21}$$

Therefore the total value of the free surface elevation at the incoming wave boundary is

$$\zeta_{total} = \zeta_{inc} + \zeta_{rad} \tag{6.22}$$

In the following, the generation and propagation of a long (non-dispersive) one-dimensional wave is presented for different physical scenarios (Koutitas, Gousidou-Koutita and Papazachos 1983). The general equation used for those simulations was taken as

$$\frac{\partial^2 \zeta}{\partial t^2} - \frac{\partial}{\partial x}\left(c_0^2 \frac{\partial \zeta}{\partial x}\right) - \lambda\frac{\partial \zeta}{\partial t} + N_b \frac{\partial}{\partial t}\left(\frac{\partial^2 \zeta}{\partial x^2}\right) = 0 \tag{6.23}$$

Example 6.1

This exercise illustrates the effects of a sloping plane bed on a one-dimensional incident linear long wave. The data provided are as follows:

> Wave amplitude of the incident wave = 1 m
> Wave period = 8 s
> Bed friction coefficient = 0.001
> Constant parameter for eddy viscosity calculation = 0.4
> Water depth at the open sea = 5 m
> Water depth at the coastline = 0.5 m
> Longitudinal length = 400 m

The discretization steps for the simulation were taken as $\Delta x = 2$ m and $\Delta t = 0.1$ s. The governing equation was Equation 6.23, thus both bed friction and turbulence energy dissipation were considered. The boundary condition for the open-sea boundary was a combination of incident and radiated waves, while for the coastline boundary it was free radiation.

The simulation was conducted for a period of 3040 seconds (38 wave periods), and the results are presented in Figure 6.2. From this figure, the wave-breaking location is easily identifiable by the sudden reduction of the wave root-mean-square value $H_{rms}^2 = \frac{1}{N}\sum_{m=1}^{N} H_m^2$, which is indicative of the turbulent energy dissipation. Notably, the estimated wave breaking location coincides with the established fact that wave breaking occurs at the point where the wave height exceeds 0.8 times the water depth ($H > 0.8h$). Another interesting observation is the smoothing of the wave breaking shape due to the diffusivity effects (fourth term in Equation 6.23). Diffusivity peaks at the point of wave breaking and then reduces.

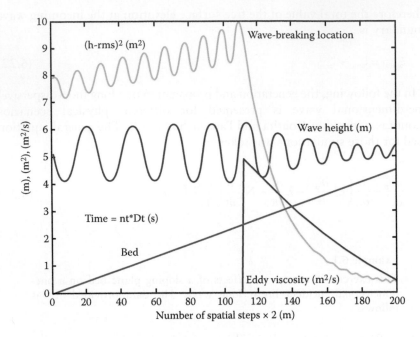

Figure 6.2 Wave shoaling over a sloping plane seafloor.

Computer code 6.1

```
% Example 6.1 One Dimensional Linear Long Wave Over a
Sloping Plane Bed
% ho = Water depth [m];
% ev = Eddy viscosity [m^2/s];
% ec = Coefficient for the eddy viscosity relation
[dimensionless];
% fb = Bed friction [s^(-1)];
% a = Incident wave amplitude [m];
% T = Wave period [seconds];
% Dx = Spatial step [m];
% Dt = Time step [s];
% nx = Number of computational steps;
% nt = Number of time steps;
clc; clear all; close all;
% Input data;
g=9.81;
ho=5;
fbo=0.001;
a=1;
T=8;
ec=0.4;
```

```
Dx=2;
Dt=0.1;
nx=200;
nt=3040;
r=g*(Dt/Dx)^2/2;
nx2=nx/2;
% Option for variable bed friction;
for i=1:nx
    fb(i)=fbo;
end
% Bathymetry data for sloping bed;
for i=1:nx
    % For horizontal bed active next line;
    % h(i)=ho;
    % For sloping bed activate next line;
    h(i)=ho-(ho-.5)*i/nx;
    bed(i)=5-h(i);
end
for i=1:nx
    z(i)=0;
    zo(i)=0;
    zm(i)=0;
end
% Estimation of the eddy viscosity;
for i=1:nx
    ev(i)=0;
    if a>h(i)/2.5
        ev(i)=ec*h(i)*sqrt(g*h(i));
    end
end
c=sqrt(g*h(1));
L=c*T;
zr1=0;
% Main program;
for k=1:nt
    % Open sea boundary conditions: Incident and back
radiated wave;
    zr2=z(2)-a*sin(2*pi*((k-1)*Dt/T-Dx/L));
    zr1=zr1+Dt/Dx*c*(zr2-zr1);
    zn(1)=a*sin(2*pi*k*Dt/T)+zr1;
    % Solution within the main domain;
    for i=2:nx-1
        br=ev(i)*Dt/
Dx^2*(z(i+1)-2*z(i)+z(i-1)-zo(i+1)+2*zo(i)-zo(i-1));

        tt=r*((h(i+1)+h(i))*(z(i+1)-z(i))-(h(i)+h(i-1))*
(z(i)-z(i-1)));
        zn(i)=2*z(i)-zo(i)+tt-Dt*fb(i)*(z(i)-zo(i))+br;
    end
```

```
    % Boundary condition at the coastline: Back radiated
wave;
    zn(nx)=z(nx)+sqrt(g*h(nx))*Dt/Dx*(z(nx-1)-z(nx));
    for i=1:nx
        zo(i)=z(i);
        z(i)=zn(i);
        % Water depth below the water surface;
        hwater(i)=z(i)+ho;
    end
    if k>nx2
        % Estimation of the variance (h-rms)^2 (Wave height
root-mean-square);
        for i=1:nx
            zm(i)=zm(i)+z(i)^2/nx;
        end
    end
end
plot(1:nx,hwater,'b','Linewidth',1.5)
hold on
plot(1:nx,zm,'g','LineWidth',1.5)
plot(1:nx,bed,'m','LineWidth',2)
plot(1:nx,ev,'r','LineWidth',1.5)
xlabel('Number of spatial steps x 2 [m]')
ylabel('[m], [m^2], [m^2/s]')
text(143,6,'Wave height [m]')
text(20,9.2,'(h-rms)^2 [m^2]')
text(20,2.7,'Time = nt*Dt [s]')
text(115,9.5,'Wave-breaking location')
text(40,1.5,'Bed')
text(112,0.5,'Eddy viscosity [m^2/s]')
```

PROBLEM 6.1

Solve the same application by making the appropriate assumptions, modifying the computer code as needed and changing one or more of the input data as suggested. In addition, you may need to change the total number of time steps (code variable 'nt').

1. Change the bed friction to f_{bo} = 0.1 s^{-1} and 0.01 s^{-1}, run the simulations and comment on the results.
2. Change the coefficient of the eddy viscosity to ec = 0, 0.2 and 1.0, then run the simulations. Compare and explain the data obtained.
3. Change the wave period to 2 s, 4 s, 16 s and 32 s. Conduct the simulations, then compare and comment on the results.
4. Change the water depth (a) from 10 m at x = 0 to 5 m at x = 400 m, and (b) from 10 m at x = 0 to 1 m at x = 400 m. Conduct the simulations, then compare and comment on the results.

5. Assume a constant water depth ($h_o = 5$ m) and wave amplitude ($a_o = 0.5$ m). Neglect the friction and turbulence effects. For the open-sea boundary apply the expression $\zeta(x,t) = 2a_o \sin (kL - \sigma t) \cos [k(x - L)]$ and for the coastal boundary use a no-flux condition (closed-end) $\dfrac{\partial \zeta}{\partial x_{end}} = 0$. Conduct the simulation and explain the physical significance of the results.

Example 6.2

This exercise deals with the effects of a submerged breakwater on the propagation of one-dimensional linear long wave. The data used are as follows:

Wave amplitude of the incident wave = 1 m
Wave period = 8 s
Bed friction coefficient = 0.001
Water depth = 5 m
Height of the breakwater = 3.33 m
Length of the breakwater = 20 m
Location of the breakwater = Mid-point of the longitudinal length
Longitudinal length = 400 m

The discretization steps for the simulation were taken as $\Delta x = 2$ m and $\Delta t = 0.1$ s. The governing equation was Equation 6.23 but the turbulent energy dissipation was neglected. The boundary condition for the open-sea boundary was a combination of incident and radiated wave, while for the coastline boundary it was free radiation.

The simulation was conducted for a period of 3200 seconds (40 wave periods) and the results are presented in Figure 6.3. From this figure, the partial wave reflection and partial wave attenuation (absorption) due to the presence of the submerged breakwater is evident. Similar conclusions can be derived from the sudden reduction of the wave height root-mean-square value $H_{rms}^2 = \dfrac{1}{N} \displaystyle\sum_{m=1}^{N} H_m^2$ at the breakwater location (Figure 6.4).

Computer code 6.2

```
% Example 6.2 One Dimensional Linear Long Wave Over a
Submerged Breakwater
% ho = Water depth [m];
% xf = Factor for sizing the height of the submerged
breakwater;
% fb = bed friction;
% a = Incident wave amplitude [m];
```

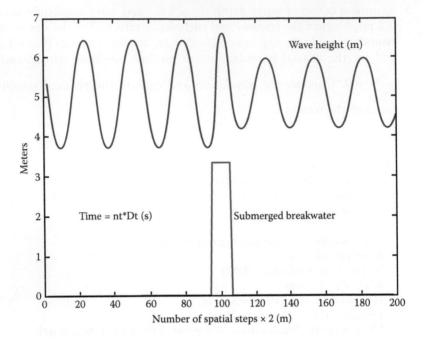

Figure 6.3 Partial wave reflection and absorption by a submerged breakwater.

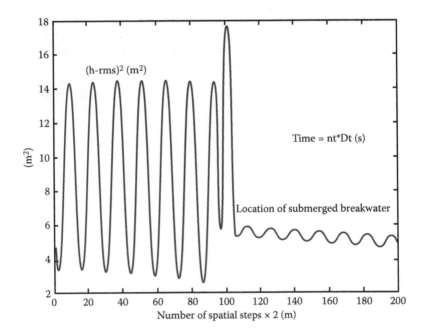

Figure 6.4 Submerged breakwater effects on the root-mean-square wave height.

```
% T = Wave period [s];
% Dx = Spatial step [m];
% Dt = Time step [s];
% nx = Number of computational steps;
% nt = Number of time steps;
clc; clear all; close all;
% Input data;
g=9.81;
ho=5;
a=1;
T=8;
xf=3;
Dx=2;
Dt=0.1;
nx=200;
nt=3200;
r=g*(Dt/Dx)^2/2;
nx2=nx/2;
% Option for variable bed friction;
for i=1:nx
    fb(i)=0.001;
end
for i=1:nx
    z(i)=0;
    zo(i)=0;
    zm(i)=0;
    h(i)=ho;
    bed(i)=0;
end
% Definition of size and location of the submerged
breakwater;
for i=nx2-5:nx2+5
    h(i)=ho/xf;
    bed(i)=-h(i)+ho;
end
c=sqrt(g*h(1));
L=c*T;
zr1=0;
% Main program;
for k=1:nt
    % Open sea boundary conditions;
    zr2=z(2)-a*sin(2*pi*((k-1)*Dt/T-Dx/L));
    zr1=zr1+Dt/Dx*c*(zr2-zr1);
    zn(1)=a*sin(2*pi*k*Dt/T)+zr1;
    % Solution of main domain;
    for i=2:nx-1
        tt=r*((h(i+1)+h(i))*(z(i+1)-z(i))-(h(i)+h(i-1))*
(z(i)-z(i-1)));
        zn(i)=2*z(i)-zo(i)+tt-Dt*fb(i)*(z(i)-zo(i));
    end
```

```
% Coastline boundary condition;
zn(nx)=z(nx)+sqrt(g*h(nx))*Dt/Dx*(z(nx-1)-z(nx));
for i=1:nx
    zo(i)=z(i);
    z(i)=zn(i);
    hdepth(i)=z(i)+ho;
end
% Estimation of the wave height variance (h-rms)^2;
if k>nx2
    for i=1:nx
        zm(i)=zm(i)+(z(i)^2)/nx;
    end
end
end
plot(1:nx,hdepth,'Linewidth',1.5)
hold on
plot(1:nx,bed,'m','LineWidth',2)
xlabel('Number of spatial steps x 2 [m]')
ylabel('Meters')
text(140,6.3,'Wave height [m]')
text(106,2,'Submerged breakwater')
text(20,2,'Time = nt*Dt [s]')
figure; plot(1:nx,zm,'','LineWidth',1.5)
hold on
xlabel('Number of spatial steps x 2 [m]')
ylabel('[m^2]')
text(20,15.2,'(h-rms)^2 [m^2]')
text(106,7,'Location of submerged breakwater')
text(140,11,'Time = nt*Dt [s]')
```

PROBLEM 6.2

Solve the same application by making the appropriate assumptions, modifying the computer code as needed and changing one or more of the input data as suggested. In addition, you may need to change the total number of time steps (code variable 'nt').

1. Change the bed friction to $f_{bo} = 0.1 \text{ s}^{-1}$ and 0.01 s^{-1}, then run the simulations and comment on the results of the two cases.
2. Change the shape of the breakwater to an orthogonal triangle of base 28 m and height 3.0 m (right angle at the right-hand side of the base). Conduct the simulation and compare the results with those of a rectangular breakwater of the same height and base.
3. Remove the breakwater and instead assume a seabed section at similar location and length having a bed friction coefficient of $f_{bo} = 0.1 \text{ s}^{-1}$ and 1 s^{-1}. Conduct the simulations, then compare and discuss the results of the two cases.

4. Change the time step to $\Delta t = 0.285$ s and $\Delta t = 0.286$ s. Conduct the simulations and explain the results.
5. Change the wave period to $T = 2$ s and 32 s. Conduct the simulations and compare the effects of the submerged breakwater on the two different period waves.

Example 6.3

This exercise demonstrates the creation of a one-dimensional linear long wave (i.e. tsunami) due to a sudden oscillation of the sea floor (i.e. earthquake) (Koutitas, Gousidou-Koutita and Papazachos 1983). The data used for the simulation are as follows:

Water depth = 4 m
Bed friction = 0.001 s^{-1}
Longitudinal length of the solution domain = 400 m
Parameter for the seismic vertical displacement = 2 m
Parameter for the seismic time occurrence = 4 s

The discretization steps for the simulation were taken as $\Delta x = 2$ m and $\Delta t = 0.1$ s. The governing equation was given by Equation 6.23, but the turbulence energy losses were neglected. The initial condition was a horizontal water surface, while both boundaries were described by a free radiation condition. The seismic oscillation was induced by a gradual 2 m rising of the seabed within a period of 4 s. The movement occurred at a seafloor segment of 20 m located at the middle of the domain.

The propagation of the resulted tsunami wave after times 12 s and 24 s is illustrated in Figure 6.5. Considering that the celerity for shallow waves is $c_o = \sqrt{gh}$, then for a water depth of 4.0 m, the tsunami front is expected to move with celerity of 6.26 m/s, a fact that is consistent with the simulation results (Figure 6.5).

Computer code 6.3

```
% Example 6.3 1-D Linear Long Wave Generated from Sudden Sea
Floor Movement
% ho = Water depth [m];
% fbo = bed friction;
% nd = Parameter of time variation of sea floor movement;
% zbmax = Parameter for maximum sea floor movement [m];
% Dx = Spatial step [m];
% Dt = Time step [s];
% nx = Number of computational steps;
% nt = Number of time steps;
clc; clear all; close all;
```

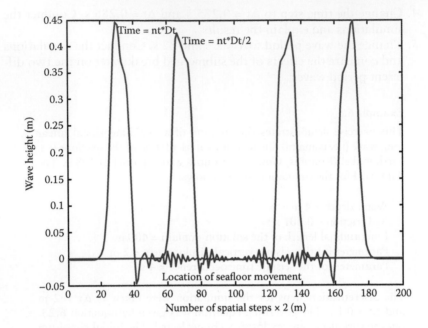

Figure 6.5 Propagation of a tsunami wave.

```
% Input data;
g=9.81;
ho=4;
fbo=0.001;
zbmax=2;
nd=4;
Dx=2;
Dt=0.1;
nx=200;
nt=240;
r=g*(Dt/Dx)^2/2;
nx2=nx/2;
% Option for variable bed friction;
for i=1:nx
    fb(i)=fbo;
end
for i=1:nx
    z(i)=0;
    zo(i)=0;
    zm(i)=0;
    zn(i)=0;
    h(i)=ho;
    bed(i)=0;
end
```

```
c=sqrt(g*h(1));
zb=0;
zbn=0;
zbo=0;
k=0;
% Main program;
for k=1:nt
    k=k+1;
    zbn=zbmax*k*Dt/nd;
    if k*Dt>nd
        zbn=zbmax;
    end
    % Movement pattern of the sea floor;
    for i=nx2-2:nx2+2
        bed(i)=(zbn-2*zb+zbo);
    end
    for i=2:nx-1
        tt=r*((h(i+1)+h(i))*(z(i+1)-z(i))-(h(i)+h(i-1))*
(z(i)-z(i-1)));
        zn(i)=2*z(i)-zo(i)+tt-Dt*fb(i)*(z(i)-zo(i))+bed(i);
    end
    zn(nx)=z(nx)+sqrt(g*h(nx))*Dt/Dx*(z(nx-1)-z(nx));
    zn(1)=z(1)+sqrt(g*h(1))*Dt/Dx*(z(2)-z(1));
    for i=1:nx
        zo(i)=z(i);
        z(i)=zn(i);
        % Wave data at half-time of the simulation period;
        % Note that z1 is recorded only if nt is an even
number;
        if k==nt/2
            for i=1:nx
                z1(i)=z(i);
            end
        end
    end
    zbo=zb;
    zb=zbn;
end
plot(1:nx,z,'','LineWidth',1.5)
hold on
plot(1:nx,z1,'r','LineWidth',1.5)
xlabel('Number of spatial steps x 2 [m]')
ylabel('Wave height [m]')
text(70,0.41,'Time=nt*Dt/2')
text(31,0.43,'Time=nt*Dt')
text(58,-0.035,'Location of sea floor movement')
```

PROBLEM 6.3

Conduct the following simulations by making the suggested changes while keeping all of the other data constant:

1. Change the longitudinal length to 4000 m, then compare and explain the simulation results after running the model for 600 and 3600 time steps.
2. Using the same seismic seafloor spatial segment, change the oscillation mode to a sudden rise of the floor to a height of 0.1 m at time t = 0.1 s followed by an additional rise to 0.2 m at t = 0.2 s and then falling back to zero elevation at t = 0.3 s. Repeat the problem using a sudden fall to a depth of −0.1 m at t = 0.1 s and an additional fall to −0.2 m at t = 0.2 s and then returning the seafloor elevation back to zero level at t = 0.3 s. Run the two simulations, and then compare and comment on the results.
3. For the right half of the solution domain, change the bed friction to $f_{bo} = 0.1$ s^{-1}. Conduct the simulation and comment on the results.
4. Assume that the water over the 20 m seismic section is 5 m deep, while the water the left of that section the is 6 m deep and to the right is 4 m deep. Run the computer program, and explain and comment on the simulation results.
5. If the anticipated seismic characteristic parameters are zbmax = 4 m and nb = 2, how much artificial bed friction is needed so that the tsunami wave does not exceed 1 m over the equilibrium mean sea level?

6.3 TWO-DIMENSIONAL LINEAR GRAVITY LONG WAVES

For a horizontal, two-dimensional domain, the linear non-dispersive long wave equation involving bed friction and wave-breaking energy losses takes the form

$$\frac{\partial^2 \zeta}{\partial t^2} = \frac{\partial}{\partial x}\left(c_o^2 \frac{\partial \zeta}{\partial x}\right) + \frac{\partial}{\partial y}\left(c_o^2 \frac{\partial \zeta}{\partial y}\right) - \lambda \frac{\partial \zeta}{\partial t} + N_b\left[\frac{\partial}{\partial t}\left(\frac{\partial^2 \zeta}{\partial x^2}\right) + \frac{\partial}{\partial t}\left(\frac{\partial^2 \zeta}{\partial y^2}\right)\right] \quad (6.24)$$

If wave dispersion effects are to be accounted for, then the equation becomes

$$\frac{\partial^2 \zeta}{\partial t^2} = \frac{\partial}{\partial x}\left(c_o c_g \frac{\partial \zeta}{\partial x}\right) + \frac{\partial}{\partial y}\left(c_o c_g \frac{\partial \zeta}{\partial y}\right) - (\sigma^2 - k^2 c_o c_g)\zeta - \lambda \frac{\partial \zeta}{\partial t}$$
$$+ N_b\left[\frac{\partial}{\partial t}\left(\frac{\partial^2 \zeta}{\partial x^2}\right) + \frac{\partial}{\partial t}\left(\frac{\partial^2 \zeta}{\partial y^2}\right)\right] \quad (6.25)$$

The preceding equations can be solved numerically by using a centred second-order finite differences scheme for both time and space derivatives. Therefore, each new value of the variable $\zeta_{i,j}^{n+1}$ at time $t + \Delta t$ is explicitly related to the neighbouring known values estimated at times t and $t - \Delta t$, or defined by either the initial or boundary conditions ($\zeta_{i,j}^{n}$, $\zeta_{i-1,j}^{n}$, $\zeta_{i+1,j}^{n}$, $\zeta_{i,j-1}^{n}$, $\zeta_{i,j+1}^{n}$, $\zeta_{i,j}^{n-1}$, $\zeta_{i-1,j}^{n-1}$, $\zeta_{i+1,j}^{n-1}$, $\zeta_{i,j-1}^{n-1}$ and $\zeta_{i,j+1}^{n-1}$). The initial conditions are usually referred to as a cold start (zero ζ values) for the first two time levels. Depending on the situation, the boundary conditions can be one of the following:

- Incident wave (open-sea) boundary, where the incoming wave, described by a sinusoidal function, interacts with the reflected and freely radiated value of ζ from the interior of the flow domain interior, similar to the one-dimensional case.
- Reflected wave, induced by reflective structural boundaries (seawalls, jetties, breakwaters, etc.). The reflection may be total or partial. For total reflection, the condition $\dfrac{\partial \zeta}{\partial n} = 0$ (n is normal to the boundary) is used, while for partial reflection this condition is modified.
- Free wave radiation conditions, where the Equation 6.21 is applied in the direction 'n' normal to the boundary.

Example 6.4

This exercise involves an incident sinusoidal, linear, non-dispersive long wave propagating perpendicularly to a detached breakwater in a coastal area of linearly decreasing water depth. The wave is partially reflected on the offshore side of the breakwater, diffracted around its ends, refracted in the area behind the breakwater, subject to shoaling and losing its energy in the breaking zone. The data used for the simulation are as follows:

> Incident wave amplitude = 0.5 m
> Incident wave period = 12 s
> Angle of incident wave = 1°
> Wave reflection coefficient = 0.5
> Parameter for the eddy viscosity relation = 0.8
> Water depth at the open-sea boundary = 8.5 m
> Water depth at the coastline boundary = 0.5 m
> Spatial domain width = 180 m
> Spatial domain length = 240 m
> Breakwater length = 90 m
> Breakwater location = 96 m for the open-sea boundary and parallel to the coastline

The governing equation was Equation 6.24 where the friction was neglected. The spatial discretization step was 3 m and the time step 0.1 s.

The program was run for 2400 time steps (20 wave cycles). The time evolution of the free surface elevation at a point located at the

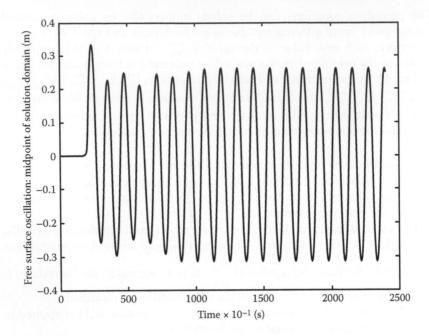

Figure 6.6 Time series of wave height at the centre point of the flow domain.

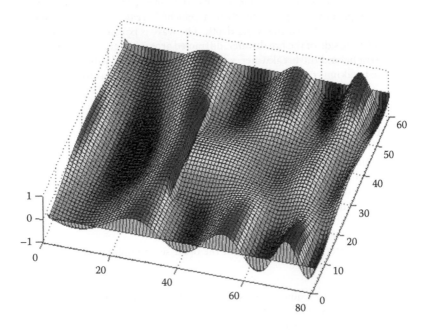

Figure 6.7 Free surface spatial distribution at time t = 2400 Δt.

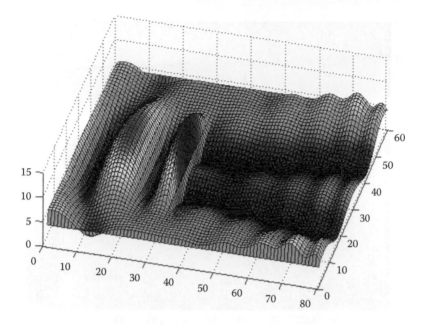

Figure 6.8 Spatial distribution of the root-mean-square wave height.

middle of the flow domain (behind the breakwater) shows that a quasi-steady periodic state is established after 10 wave periods, resulting in a reduction of the wave amplitude by half (Figure 6.6). The impact of the breakwater and the free wave radiation boundary conditions on the wave propagation are captured in the three-dimensional spatial distribution of the free surface elevation taken at time t = 2400 Δt (Figure 6.7).

In addition, computation of the root-mean-square wave height and its three-dimensional rendering throughout the solution domain effectively demonstrates the wave reflection, refraction, diffraction and shoaling processes (Figure 6.8). This type of result can be used as a design and operational tool, supporting decision-making regarding the efficiency of the breakwater structure. Finally, the instantaneous wave-induced velocities pattern at time t = 2400 Δt, estimated on the basis of Equation 6.14 extended in two dimensions, is presented in Figure 6.9. Those velocities were used for the estimation of the radiation stresses and the consequent wave-induced circulation (in an averaged-over-the-period sense) patterns.

Computer code 6.4

```
% Example 6.4 Two Dimensional Long Wave with Reflecting
Breakwater
% a = Wave amplitude [m];
```

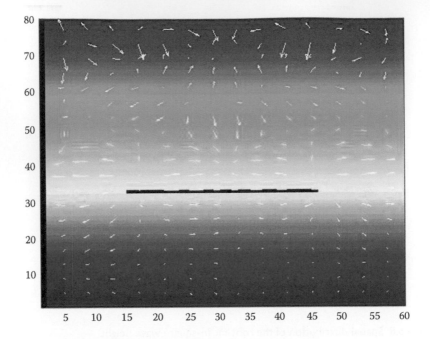

Figure 6.9 Wave-generated current pattern at time t = 2400 Δt.

```
% T = Wave period [T];
% theta = Incident wave angle [degrees];
% rf = Reflection coefficient (rf = 1 for full reflection);
% ec = Parameter for the eddy viscosity;
% Dd = Spatial step in both x and y directions [m];
% ho = Water depth at the open sea boundary;
% Dt = Time step [s];
% nx = Number of computational steps in the x-direction;
% ny = Number of computational steps in the y-direction;
% nt = Number of computational time steps;
clc; clear all, close all;
% input data;
g=9.81;
a=0.5;
T=12;
theta=1;
rf=0.5;
ec=0.8;
ho=8.5;
Dd=3;
Dt=0.1;
nx=60;
ny=80;
nt=2400;
```

```
% Initialization of variable;
for i=1:nx
    for j=1:ny
        h(i,j)=0;
        z(i,j)=0;
        zo(i,j)=0;
        u(i,j)=0;
        v(i,j)=0;
        sxx(i,j)=0;
        syy(i,j)=0;
        sxy(i,j)=0;
        hrms(i,j)=0;
    end
end
ar=0;
al=0;
ao=0;
au=0;
chk=0;
% Water depth data;
for j=1:ny
    for i=2:nx-1
        h(i,j)=ho-j/ny*8;
    end
end
% Definition of the breakwater location and geometry;
for i=nx/4:3*nx/4
    h(i,ny/2.5)=0;
end
c1=sqrt(g*h(2,1));
b=theta*pi/180;
elo=c1*T/sin(b);
phi=2*pi*Dd/elo;
% Estimation of eddy coefficient in wave breaking zone;
for j=1:ny
    for i=1:nx
        if h(i,j)> 2.5*a
            ed(i,j)=0;
        else
            ed(i,j)=ec*h(i,j)*sqrt(g*h(i,j));
        end
    end
end
% Main program;
for k=1:nt
    % Open sea boundary condition: Incident and radiating
waves;
    for i=1:nx
        zr1=z(i,1)-a*sin(2*pi*((k-1)*Dt/T)+(i-1)*phi);
        zr2=z(i,2)-a*sin(2*pi*((k-1)*Dt-Dd/c1)/T+(i-1)*phi);
```

```
            zr1=zr1+Dt/Dd*c1*(zr2-zr1);
            zn(i,1)=zr1+a*sin(2*pi*k*Dt/T+(i-1)*phi);
    end
    % Solution over the flow domain;
    for j=2:ny-1
        for i=2:nx-1
            if h(i,j)> 0.1
                al=z(i-1,j);
                chk=1000;
            end
            if chk==1000
                if h(i-1,j)< 0.1
                    al=z(i,j)*rf+(1-rf)*zo(i,j);
                end
                ar=z(i+1,j);
                if h(i+1,j)< 0.1
                    ar=z(i,j)*rf+(1-rf)*zo(i,j);
                end
                ao=z(i,j+1);
                if h(i,j+1)< 0.1
                    ao=z(i,j)*rf+(1-rf)*zo(i,j);
                end
                au=z(i,j-1);
                if h(i,j-1)< 0.1
                    au=z(i,j)*rf+(1-rf)*zo(i,j);
                end
                hr=(h(i+1,j)+h(i,j))/2;
                hl=(h(i-1,j)+h(i,j))/2;
                ho=(h(i,j+1)+h(i,j))/2;
                hu=(h(i,j)+h(i,j-1))/2;
                zn(i,j)=z(i,j)*2-zo(i,j)+(Dt/Dd)^2*g*(hr*
(ar-z(i,j))-hl*(z(i,j)-al)+ho*(ao-z(i,j))-hu*(z(i,j)-au));
                zn(i,j)=zn(i,j)+ed(i,j)*Dt/
Dd^2*(z(i+1,j)+z(i-1,j)+z(i,j+1)+z(i,j-1)-4*z(i,j)-
zo(i+1,j)-zo(i-1,j)-zo(i,j+1)-zo(i,j-1)+4*zo(i,j));
                if abs(zn(i,j)) < 0.00001
                    zn(i,j)=0;
                end
            end
            chk=0;
        end
    end
    % Coastline boundary condition;
    for i=2:nx-1
        if h(i,ny)> 0.1
            zn(i,ny)=z(i,ny)+Dt/Dd*sqrt(g*h(i,ny))*(z(i,ny-
1)-z(i,ny));
        end
    end
```

```
    % Left and right boundary conditions: Reflection or free
radiation;
    for j=1:ny
        if h(1,j)>0.1
            % Reflection;
            % zn(1,j)=zn(2,j);
            % Free radiation;
            zn(1,j)=z(1,j)+Dt/Dd*sqrt(g*h(1,j))*(z(2,j)-z
(1,j));
        end
        if h(nx,j)> 0.1
            % zn(nx,j)=zn(nx-1,j);
            zn(nx,j)=z(nx,j)+Dt/Dd*sqrt(g*h(nx,j))*(z(nx-
1,j)-z(nx,j));
        end
    end
    % Updating the z wave values;
    for j=1:ny
        for i=1:nx
            zo(i,j)=z(i,j);
            z(i,j)=zn(i,j);
        end
    end
    % Free surface time series at the middle of the field;
    zmid(k)=z(nx/2,ny/2);
    if k>nt/2
        % Estimation of flow velocity field and radiation
stresses;
        for j=2:ny-1
            for i=2:nx-1
                if h(i,j)>0.1 && h(i-1,j)>0.1 && h(i+1,j)>0.1
                    u(i,j)=u(i,j)-Dt/Dd*g*(z(i+1,j)-z
(i-1,j))/2;
                end
            end
        end
        for j=2:ny-1
            for i=2:nx-1
                if h(i,j)>0.1 && h(i,j-1)>0.1 && h(i,j+1)>0.1
                    v(i,j)=v(i,j)-Dt/Dd*g*(z(i,j+1)-z
(i,j-1))/2;
                end
            end
        end
        for j=2:ny-1
            for i=2:nx-1
                if h(i,j)>0.1
```

```
                              sxx(i,j)=sxx(i,j)+h(i,j)*u(i,j)^2+g/2*z
(i,j)^2;
                              syy(i,j)=syy(i,j)+h(i,j)*v(i,j)^2+g/2*z
(i,j)^2;
                              sxy(i,j)=sxy(i,j)+h(i,j)*u(i,j)*v(i,j);
                    end
                end
            end
            % Computation of the rms wave height;
            for j=1:ny
                for i=1:nx
                    hrms(i,j)=hrms(i,j)+z(i,j)^2;
                end
            end
        end
        index=k
end
for j=1:ny
    for i=1:nx
        hrms(i,j)=sqrt(hrms(i,j)/nx*2)*2;
        if h(i,j)<0.1
            hrms(i,j)=4;
        end
    end
end
plot(1:nt,zmid,'k','Linewidth',1.5)
xlabel('Time x 10^-^1 [s]')
ylabel('Free-surface oscillation : Midpoint of solution
domain [m]')
figure
i=1:nx;
j=1:ny;
fplot=z(i,j);
fplot1=hrms(i,j);
surf(j,i,fplot)
figure
surf(j,i,fplot1)
figure
[i,j]=meshgrid(1:1:nx,1:1:ny);
pcolor(i,j,h'); hold on;
%colormap(gray(2))
shading flat
% Plotting at every node;
% quiver(i,j,u',v')
% Plotting at every four points;
quiver(i(1:4:end,1:4:end),j(1:4:end,1:4:end),(u(1:4:end,1:4:
end))',(v(1:4:end,1:4:end))','Color','w'); hold on;
```

PROBLEM 6.4

Make the suggested changes in the computer code while keeping the rest of the data constant:

1. Change the reflection coefficient to zero (absorption) and one (full reflection). Conduct the simulations, and then compare and comment on the results.
2. Change the eddy viscosity parameter from 0.8 to 0.2 and 1.2. In addition, change the right and left boundary conditions to reflection instead of free radiation. Conduct the simulations, and then compare and comment on the results.
3. Remove a 30 m central section of the breakwater leaving two sections of 30 m each. Conduct the simulation and comment on the changes of the wave and current pattern.
4. Make the water depth constant and equal to 6 metres. Run the simulation for three different breakwater reflection scenarios (rf = 0, 0.5 and 1) and comment on the results.
5. Run the simulation by changing the direction of wave incidence to 30°. Then change the lateral size of the solution domain from 180 m to 1440 m and run the simulation again. Compare and comment on the results obtained from the two simulations.

6.4 HARBOUR BASIN RESONANCE

An important operational problem pertaining to certain scale-size water basins is the investigation of the eigen periods and eigenmodes of oscillation and the avoidance of resonance. Eigen periods are the natural periods of oscillations of water bodies. The phenomenon of resonance refers to a substantial increase of the amplitude of oscillation of the basin water masses that have the tendency to oscillate according to natural periods of oscillation, when they are triggered by an external periodic pulse having the same period as one of the natural periods. The recognition of those periods is important in the design of harbours in order to avoid the resonance phenomenon caused by long incident waves generating and propagating from the open sea to the harbour entrance. It is also important in the analysis of storm surges, that is, the meteorologically induced long waves on the water masses of coastal basins or lakes.

The system of equations that can be used to model the phenomenon of basin resonance is the one developed for gravity long waves. This system comprises of Equation 6.13 and Equation 5.24 re-written as

$$\frac{\partial u}{\partial t} + u\frac{\partial u}{\partial x} = -g\frac{\partial \zeta}{\partial x} - f_b\frac{u|u|}{h} \tag{6.26}$$

where ζ is the oscillation of the free surface, h is the static water depth, and f_b is a dimensionless bed friction coefficient.

The phenomenon can be analysed by monitoring the excitation of the basin water that caused a localized initial impulse, that is, a sudden increase of the free surface at one location of the basin. This local perturbation, in the form of a Dirac spike, mobilizes the water masses and generates a complex wave pattern, as the initial perturbation contains a large number of Fourier components, and the energy passes from one period to another while affected by the non-linear terms. The time series of the free surface elevation ζ_n (for a time step Δt) at a location (the same with the impulse location or another) is analysed using a Fourier series of sine and cosine components. The local maxima of the amplitude of the various Fourier components in the periodogram reveal the natural periods of oscillation.

The signal $\zeta(t)$ for a time duration of $T = n\Delta t$ seconds is analysed and the coefficients α_k and β_k of the sine and cosine Fourier components are calculated by using the classical formulas

$$\alpha_k = \frac{2}{N} \sum_{n=1}^{N} \zeta_n \sin\left(\frac{2kn\Delta t}{T} \right) \tag{6.27}$$

$$\beta_k = \frac{2}{N} \sum_{n=1}^{N} \zeta_n \cos\left(\frac{2kn\Delta t}{T} \right) \tag{6.28}$$

where n = 1 to N is the number of time steps, and k = 1 to M is the number of the Fourier components. Then the amplitude of the Fourier component $\sqrt{\alpha_k^2 + \beta_k^2}$ is plotted against k, and for each k_j value, where the amplitude is maximized, the eigen period T_j is estimated as

$$T_j = \frac{T}{k_j} \quad \text{for } j = 0, 1, 2, 3,... \tag{6.29}$$

For the case of constant depth orthogonal basins with one open-sea or closed-end boundary, the eigen periods can be estimated analytically by the Merian formula as

$$T_j = \frac{2L_B}{(j+1)\sqrt{gh}} \quad \text{for } j = 0, 1, 2, 3, ... \tag{6.30}$$

where L_B is the length of the basin and h is the undisturbed water depth.

Example 6.5

This exercise considers a one-dimensional basin subject to a sudden excitation (Dirac function) of the water surface at a certain location. The data utilized for the simulation and the Fourier series analysis are as follows:

Basin length = 1000 m
Undisturbed water depth = 5 m
Excitation amplitude = 0.5 m
Bed friction coefficient = 0.001
Number of Fourier series components = 40

The excitation occurred at a distance of 980 m. Data were collected at two points, one at a distance of 200 m and another at a distance of 500 m (mid-point). The spatial step was taken as 10 m while the time step as 0.1 s. The simulation was conducted for a period of 1000 seconds.

The time series of the free surface oscillation at the mid-point of the basin is presented in Figure 6.10. From this figure, the primary eigen period of natural oscillation (T_o) of about 275 seconds is easily recognizable. From the same figure, the attenuation in time of the oscillation due to frictional and other effects is also evident. It should be emphasized that the evolution of the various eigen periods depends on the type of excitation (i.e. Dirac function in this example).

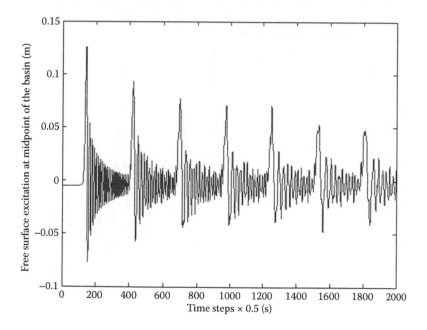

Figure 6.10 Free surface excitation at the midpoint of the basin.

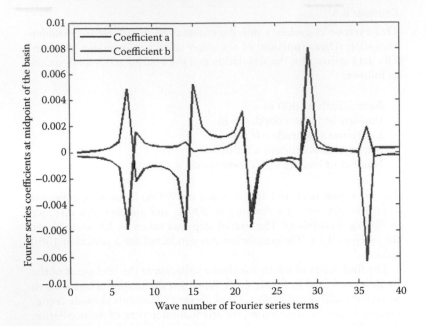

Figure 6.11 Fourier series α_k and β_k coefficients at the midpoint of the basin.

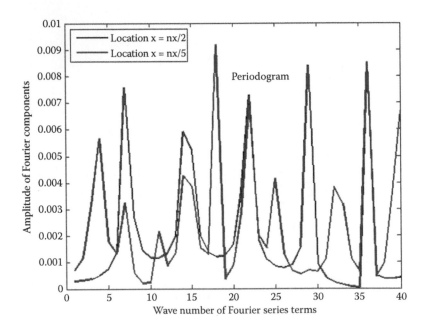

Figure 6.12 Free surface Fourier series periodogram.

Table 6.1 Natural periods estimated by the Fourier analysis and the Merian formula

Period mode, j	k_j value (Figure 6.12)	Natural period, T_{jM} (Merian formula)	Natural period, T_{js} (simulation results)
0	4	286	250
1	7	143	143
2	11	95	91
3	14	71	71
4	18	57	56
5	22	48	45

The variability of all 40 Fourier series coefficients α_k and β_k at the mid-point of the basin is depicted in Figure 6.11.

Finally, from the periodogram, that is, plot of the amplitude $\sqrt{\alpha_k^2 + \beta_k^2}$ versus the wave number k (= 1 to 40) of the Fourier series components (Figure 6.12), the corresponding k_j wave numbers are identified and the periods of natural oscillations are estimated according to Equation 6.28. The natural oscillation periods estimated by the Fourier series analysis and by the Merian formula (Equation 6.29) are listed in Table 6.1.

Computer code 6.5

```
% Example 6.5 Impulse Response of a Harbor Basin
% ho = Water depth [m];
% zo = Excitation amplitude [m];
% fb = Bed friction [dimensionless];
% fsero = Number of Fourier series terms;
% Dx = Computational spatial step [m];
% Dt = Computational time step [s];
% nx = Number of spatial steps;
% nt = Number of time steps;
clc; clear all; close all;
% Input data;
g=9.81;
ho=5;
zo=.5;
fb=0.001;
fsero=40;
Dx=10;
Dt=0.5;
nx=100;
nt=2000;
% Initialization of variables;
for i=1:nx
    z(i)=0;
```

```
    h(i)=0;
    u(i)=0;
    un(i)=0;
end
% Definition of constant depth;
for i=2:nx-2
    h(i)=ho;
end
z(nx-2)=zo;
% Main program;
for k=1:nt
    for i=2:nx-2
        z(i)=z(i)-Dt/
Dx*((h(i)+h(i+1))*u(i+1)-(h(i)+h(i-1))*u(i))/2;
    end
    % Definition of the spatial location to be analyzed;
    % Note that nx/2 and nx/5 must be integers;
    x(k)=z(nx/2);
    x1(k)=z(nx/5);
    for i=3:nx-2
        if u(i)<0
            adv=u(i)*(u(i+1)-u(i))/Dx*Dt;
        else
            adv=u(i)*(u(i)-u(i-1))/Dx*Dt;
        end
        un(i)=u(i)-adv-Dt/Dx*g*(z(i)-z(i-1))-
Dt*fb*u(i)*abs(u(i))/(h(i)+h(i+1))*2;
    end
    if h(nx-1)>0.1
        un(nx-1)=z(nx-2)*sqrt(g/h(nx-2));
    end
    for i=2:nx-1
        u(i)=un(i);
    end
end
% Fourier series analysis at two selected spatial locations;
fser=fsero;
av=0;
av1=0;
for k=1:nt
    av=av+x(k);
    av1=av1+x1(k);
end
av=av/nt;
av1=av1/nt;
for k=1:nt
    x(k)=x(k)-av;
    x1(k)=x1(k)-av1;
end
% Fourier coefficient a;
```

```
for m=1:fser
    a(m)=0;
    a1(m)=0;
    for k=1:nt
        a(m)=a(m)+x(k)*sin(2*pi*m*(k-1)/nt);
        a1(m)=a1(m)+x1(k)*sin(2*pi*m*(k-1)/nt);
    end
    a(m)=a(m)*2/nt;
    a1(m)=a1(m)*2/nt;
end
% Fourier coefficient b;
for m=1:fser
    b(m)=0;
    b1(m)=0;
    for k=1:nt
        b(m)=b(m)+x(k)*cos(2*pi*m*(k-1)/nt);
        b1(m)=b1(m)+x1(k)*cos(2*pi*m*(k-1)/nt);
    end
    b(m)=b(m)*2/nt;
    b1(m)=b1(m)*2/nt;
end
amax=0;
% Fourier amplitude periodogram;
for m=1:fser
    ampl(m)=sqrt(a(m)^2+b(m)^2);
    ampl1(m)=sqrt(a1(m)^2+b1(m)^2);
end
plot(1:nt,x,'b','Linewidth',1.0)
xlabel('Time steps x 0.5 [s]');
ylabel('Free surface excitation at mid-point of the basin
[m]')
figure
plot(1:fser,a,'b','Linewidth',1.5);
hold on
plot(1:fser,b,'r','Linewidth',1.5);
xlabel('Wave number of Fourier series terms');
ylabel('Fourier series coefficients at mid-point of the
basin');
legend('Coefficient a','Coefficient b',2);
figure
plot(1:fser,ampl,'b','Linewidth',1.5);
hold on
plot(1:fser,ampl1,'r','Linewidth',1.5);
xlabel('Wave number of Fourier series terms');
ylabel('Amplitude of Fourier components');
legend('Location x=nx/2','Location x=nx/5',2);
text(20,0.0080,'Periodogram')
```

PROBLEM 6.5

Make the following suggested changes to Computer code 6.5 while keeping the rest of the data constant:

1. Change the friction coefficient to 0.01, 0.1 and 1.0. Run the simulations, and then compare and comment on your results.
2. Create an excitation of 0.5 m at point nx/4 and then at point 3nx/4. Run the simulations, then compare the results. Also investigate the scenario where a simultaneous excitation occurs at both of the aforementioned points.
3. Assume a water depth that varies linearly from 4 m at the ends of the basin to 8 m at the centre. Then change the configuration, setting the depth to 8 m at the ends and 4 m at the centre. Run the simulations and comment on how the phenomenon is affected by the change in the shape of the bed.
4. Change the number of spatial steps to nx = 250 and 500. Run the simulations, then comment on the results.
5. Place a submerged breakwater of 1 m height at the midsection of the basin. Conduct the simulation and compare the results at the same point without the breakwater.

Chapter 7

Flow in porous media

7.1 GENERAL GROUNDWATER FLOW EQUATIONS

Porous media flows belong to the general class of creeping flows, those with very small velocities. There is a wide variety of significant engineering, ecological and geological applications that require an understanding of flow through porous media. Some examples include flow in confined and unconfined aquifers, groundwater contamination, saltwater intrusion, management of well-field withdrawal or recharge rates, and so forth. Most of the groundwater phenomena can be described by partial differential equations mainly of the parabolic or elliptic type, along with the appropriate boundary and initial conditions (Bear and Verruijt 1987).

7.1.1 Darcy equation

The most significant concept in groundwater flow is described by the experimentally derived Darcy equation:

$$\vec{U} = -\tilde{K}\nabla h = \nabla\varphi \qquad (7.1)$$

where $\vec{U}$ is the seepage or Darcy velocity within the porous medium, h is the piezometric head, $\varphi = -\tilde{K}h$ is the velocity potential, and $\tilde{K}$ is the second-order symmetric permeability (or hydraulic conductivity) tensor defined as

$$\tilde{K} = \begin{bmatrix} K_{xx} & K_{xy} & K_{xz} \\ K_{yx} & K_{yy} & K_{yz} \\ K_{zx} & K_{zy} & K_{zz} \end{bmatrix} \qquad (7.2)$$

where K_{ij} are the permeability coefficients with respect to a Cartesian coordinate system. By aligning the orthogonal system along the principal axes

of anisotropy x′, y′, z′, only the diagonal terms of the tensor remain different than zero. For a homogeneous isotropic aquifer, the tensor is reduced
to a single scalar permeability coefficient, K (= $K_{xx} = K_{yy} = K_{zz}$). From the
Darcy equation it is evident that the piezometric gradient is linearly related
to the velocity, which is an experimental confirmation that the flow is laminar. The Darcy velocity estimated by Equation 7.1 differs from the actual
flow velocity through the pores $\vec{V}$. By defining the total porosity (n) as the
percentage of the volume of voids over the total volume of the sample, and
by specifying the effective porosity (n_e) as the percentage of interconnected
pore space available for groundwater movement ($n_e < n < 1$), the relationship between the actual and Darcy velocities is

$$\vec{V} = \frac{\vec{U}}{n_e} \tag{7.3}$$

The permeability coefficient (K) is related to the permeability (or intrinsic
permeability; k) through the relationship

$$K = \frac{\rho g k}{\mu} \tag{7.4}$$

where ρ is the density and μ is the dynamic viscosity. From Equation 7.4 it is
evident that the intrinsic permeability, k, with area units depends solely on
the soil matrix, while the permeability coefficient, K, with velocity units,
depends both on the soil matrix and the fluid properties.

7.1.2 General form of the continuity equation

The most general continuity equation including unsteady compressible
flows, written in differential format, is given as

$$\frac{\partial \rho}{\partial t} + \nabla \cdot (\rho \vec{U}) = 0 \tag{7.5}$$

or

$$\frac{d\rho}{dt} + \rho \nabla \cdot \vec{U} = 0 \tag{7.6}$$

For the case of groundwater flows, Equation 7.5 can be re-written more
precisely as

$$\frac{\partial (\langle \rho \rangle \langle n_e \rangle)}{\partial t} + \nabla \cdot \left(\langle \rho \rangle \langle \vec{U} \rangle \right) \pm \langle q \rangle = 0 \tag{7.7}$$

where q is the discharge from withdrawal (+) or injection (−) wells, and < > denote spatially averaged quantities. In case of incompressible fluids (ρ is constant) and assuming a constant effective porosity, the continuity equation reduces to

$$\nabla \cdot \vec{U} = \text{div}\vec{U} = \frac{\partial u}{\partial x} + \frac{\partial v}{\partial y} + \frac{\partial w}{\partial z} = 0 \tag{7.8}$$

Combining any of the preceding continuity equations along with the Darcy equation (Equation 7.1) can lead to the derivation of the particular forms of groundwater flow equations (de Marsily 1986). It should be noted that the Darcy equation serves as a reduced form of the momentum equation.

7.1.3 Flow in confined aquifers

In confined aquifers, due to the high pressures induced, compressibility effects cannot be neglected. Therefore, both the pore-volume (soil matrix) compressibility along the vertical z-direction, α_p, and the water compressibility, β, are accounted for. More specifically these compressibility parameters are estimated as

$$\alpha_p = \frac{1}{n\Delta z} \frac{d(n\Delta z)}{dp_w} \tag{7.9}$$

$$\beta = -\frac{1}{V_w} \frac{dV_w}{dp_w} \tag{7.10}$$

where p_w is the water pressure and V_w is the water volume. The compressibility effects are combined within the specific storage, S_s, defined as

$$S_s = \rho g n(\alpha_p + \beta) = \frac{1}{V_w} \frac{dV_w}{dh} \tag{7.11}$$

where h is the piezometric head. Combining the continuity and momentum equations and accounting for compressibility effects, the general equation for flow in a confined aquifer reads

$$S_s \frac{\partial h}{\partial t} = \frac{\partial}{\partial x}\left(K_x \frac{\partial h}{\partial x}\right) + \frac{\partial}{\partial y}\left(K_y \frac{\partial h}{\partial y}\right) + \frac{\partial}{\partial z}\left(K_z \frac{\partial h}{\partial z}\right) \tag{7.12}$$

Equation 7.12 is a linear partial differential equation (PDE) of the parabolic type that describes the spatial and temporal distributions of the piezometric head in a three-dimensional, confined, anisotropic and inhomogeneous saturated aquifer. For a homogeneous but anisotropic aquifer, Equation 7.12 becomes

$$S_s \frac{\partial h}{\partial t} = K_x \frac{\partial^2 h}{\partial x^2} + K_y \frac{\partial^2 h}{\partial y^2} + K_z \frac{\partial^2 h}{\partial z^2} \qquad (7.13)$$

while for a homogeneous and isotropic aquifer it reduces to

$$\frac{S_s}{K} \frac{\partial h}{\partial t} = \frac{\partial^2 h}{\partial x^2} + \frac{\partial^2 h}{\partial y^2} + \frac{\partial^2 h}{\partial z^2} \qquad (7.14)$$

Under steady-state conditions Equation 7.14 reduces to the Laplace equation written in terms of the piezometric head, h,

$$\nabla^2 h = \frac{\partial^2 h}{\partial x^2} + \frac{\partial^2 h}{\partial y^2} + \frac{\partial^2 h}{\partial z^2} = 0 \qquad (7.15)$$

or the velocity potential, φ,

$$\nabla^2 \varphi = \frac{\partial^2 \varphi}{\partial x^2} + \frac{\partial^2 \varphi}{\partial y^2} + \frac{\partial^2 \varphi}{\partial z^2} = 0 \qquad (7.16)$$

For a two-dimensional horizontal aquifer of average thickness b, Equation 7.14 is known as the linear Boussinesq equation and reads

$$\frac{S}{T} \frac{\partial h}{\partial t} = \frac{\partial^2 h}{\partial x^2} + \frac{\partial^2 h}{\partial y^2} \qquad (7.17)$$

where $T = bK$ is the transmissivity and $S = S_s b$. By defining the drawdown $s = h_o - h$, where h_o is a reference piezometric head, the Boussinesq equation in terms of the drawdown becomes

$$\frac{S}{T} \frac{\partial s}{\partial t} = \frac{\partial^2 s}{\partial x^2} + \frac{\partial^2 s}{\partial y^2} \qquad (7.18)$$

7.1.4 Flow in unconfined aquifers

In unconfined (free surface) aquifers, the deformation of the soil matrix and the water compressibility are negligible as compared to the changes of the water table. Therefore the specific storage $S_s = 0$ and Equation 7.12 reduces to

$$\frac{\partial}{\partial x}\left(K_x \frac{\partial h}{\partial x}\right) + \frac{\partial}{\partial y}\left(K_y \frac{\partial h}{\partial y}\right) + \frac{\partial}{\partial z}\left(K_z \frac{\partial h}{\partial z}\right) = 0 \tag{7.19}$$

The difficulty in solving Equation 7.19 lies on the treatment of the free surface boundary. This issue can be treated by applying mass balance on a control volume over an impermeable bed where the fluxes through the sides balance the rise or fall of the water table. By using the Dupuit approximation (small water-surface gradients), the horizontal velocities along the thickness of the aquifer can be approximated as

$$U_x = -K_x \frac{\partial h}{\partial x} \tag{7.20}$$

$$U_y = -K_y \frac{\partial h}{\partial y} \tag{7.21}$$

In addition, by introducing the 'apparent specific yield', S_{ya}, as the ratio between the volume of water added (or removed) from the saturated aquifer over the change in volume of the aquifer below the water table, Equation 7.19 can be modified as

$$S_{ya} \frac{\partial h}{\partial t} = \frac{\partial}{\partial x}\left(K_x h \frac{\partial h}{\partial x}\right) + \frac{\partial}{\partial y}\left(K_y h \frac{\partial h}{\partial y}\right) \tag{7.22}$$

where h is the water depth below the water table. Equation 7.22 is a non-linear PDE of the parabolic type which applies to anisotropic and inhomogeneous unconfined saturated aquifers. Under homogeneous and isotropic conditions, Equation 7.22 becomes the non-linear Boussinesq equation:

$$\frac{S_{ya}}{K} \frac{\partial h}{\partial t} = \frac{\partial}{\partial x}\left(h \frac{\partial h}{\partial x}\right) + \frac{\partial}{\partial y}\left(h \frac{\partial h}{\partial y}\right) \tag{7.23}$$

The Boussinesq equation can be linearized by expanding the nonlinear terms and neglecting the second-order terms:

$$\frac{\partial}{\partial x}\left(h \frac{\partial h}{\partial x}\right) = \left(\frac{\partial h}{\partial x}\right)^2 + h \frac{\partial^2 h}{\partial x^2} \approx h \frac{\partial^2 h}{\partial x^2} \approx h_o \frac{\partial^2 h}{\partial x^2} \tag{7.24}$$

$$\frac{\partial}{\partial y}\left(h\frac{\partial h}{\partial y}\right)=\left(\frac{\partial h}{\partial y}\right)^{2}+h\frac{\partial^{2}h}{\partial y^{2}}\approx h\frac{\partial^{2}h}{\partial y^{2}}\approx h_{o}\frac{\partial^{2}h}{\partial y^{2}} \qquad (7.25)$$

where h_o is the average water depth below the water table. Then the linearized form of the Boussinesq equation (Equation 7.26) is identical to Equation 7.17, where the specific storage has been replaced by the apparent specific yield:

$$\frac{S_{ya}}{h_{o}K}\frac{\partial h}{\partial t}=\frac{\partial^{2}h}{\partial x^{2}}+\frac{\partial^{2}h}{\partial y^{2}} \qquad (7.26)$$

7.1.5 Boundary conditions in groundwater flow domains

The boundary conditions necessary to complete the mathematical model can be described in terms of the velocity potential or the piezometric head. The various types of boundary conditions can be effectively illustrated by considering flow through an earthen embankment (Figure 7.1).

- Impervious boundary (surface HAGI): Since the normal-to-the-boundary seepage velocity is zero, then $\frac{\partial\varphi}{\partial n}=0$, where n is the unit vector normal to the boundary.
- Constant head (reservoir or stream) boundary (surface AB and FG): The value of φ is constant, related to the elevation of the reservoir free surface with respect to a reference datum. For instance, at surface AB

  $$\varphi=-Kh=-K\left(\frac{p_{1}}{\rho g}+z_{1}\right)=-K(y_{1}+z_{1})=-K(y_{o}+z_{o})=\text{constant}.$$

- Seepage boundary (surface EF): The water outflows from the pores to the air. On that boundary the potential is $\varphi=-Kh=-K\left(\frac{p_{atm}}{\rho g}+z\right)\Rightarrow\varphi+Kz=-K\frac{p_{atm}}{\rho g}=\text{constant}$. Thus, the potential varies linearly with the elevation z, measured from the reference datum.
- On a free surface boundary (surface BE), in the case of unconfined flows, the surface geometry is described by the function h(x,y,t). This surface is a stream line, where the Bernoulli equation describes the conservation of energy in linearized form as $\varphi=-Kh$. This dynamic condition is complemented by a kinematic condition, which after linearization, takes the form which defines the free surface $\frac{\partial h}{\partial t}=-\frac{1}{n_{e}}\frac{\partial\varphi}{\partial z}$.

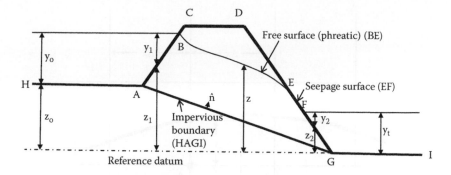

Figure 7.1 Types of boundary conditions in groundwater flow.

7.2 NUMERICAL SOLUTIONS OF GROUNDWATER FLOW APPLICATIONS

7.2.1 Steady-state vertical two-dimensional groundwater flows

In the case of vertical two-dimensional steady-state flow, the governing equation is the Laplace equation. By using a four-point finite differences scheme, the discretized equation in terms of the velocity potential is similar to that of Equation 3.16 (see Chapter 3):

$$\varphi_{i,j} = \frac{1}{4}\left(\varphi_{i+1,j} + \varphi_{i-1,j} + \varphi_{i,j+1} + \varphi_{i,j-1}\right) \tag{7.27}$$

Equation 7.27 can be solved very efficiently by using an iterative 'relaxation' procedure such as the Gauss-Seidel method. For confined flows, the boundary conditions are fixed and predetermined. For unconfined flows, the geometry of the water table has to be determined using the free surface boundary condition. The discretization of the free surface boundary can be accomplished by means of an explicit forward finite differences scheme, according to the notations shown in Figure 7.2, as

$$\frac{h_{i,j}^{n+1} - h_{i,j}^{n}}{\Delta t} = -\frac{K}{n_e}\frac{h_{i,s}^{n} - h_{i,j}^{n}}{\Delta s} = -\frac{1}{n_e}\frac{\varphi_{i,s}^{n} - \varphi_{i,j}^{n}}{\Delta s} \tag{7.28}$$

where n is the time iteration number.

Therefore, for unconfined aquifers the solution involves an elliptic equation for the interior domain (Equation 7.27) and a hyperbolic equation for the free surface (Equation 7.28). On the lateral solid boundaries, either the

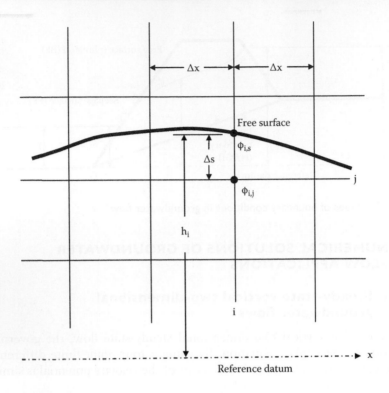

Figure 7.2 Discretization of the free surface boundary.

normal derivatives are properly modified, to conform to non-uniform grids, or the domain is extended to a node outside the flow domain, where the φ values are extrapolated from the interior of the flow domain.

Example 7.1

The problem to be dealt with involves groundwater flow below a dam with a sheet pile. The solution domain comprises of an aquifer of known geometry, and the goal is to simulate the distribution of the velocity potential $\varphi(x, y)$ or the piezometric head, h. That information can help to estimate the seepage losses (water escaping under the dam) and to assess the danger of downstream soil destabilization (piping). Also, since the zone under the dam usually is not perfectly sealed and is semi-permeable, the flow under the dam needs to be investigated (Figure 7.3).

The mathematical model consists of the Laplace equation in its finite differences form (Equation 7.27). The discretization of the flow domain is done using a rectangular grid. Boundaries of constant pressure head are the soil surface upstream and downstream of the dam. Impermeable boundaries are the dam base (if sealed), the aquifer bed

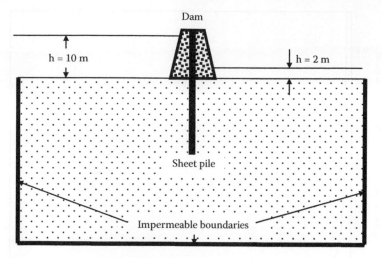

Figure 7.3 Flow under a dam with a sheet pile.

and, the far upstream and downstream limits of the solution domain. The nodes of the two-dimensional grid are characterized by an auxiliary matrix $M(i,j)$ (the i index along the x-axis and the j index along the y-axis). The values given to the elements of the matrix are $M = 0$ for the interior flow domain, $M = 1$ on impermeable boundaries and M = any real number different than 1, signifying the magnitude of any constant head boundary in metres. Other data used for this application are as follows:

Upstream piezometric head = 10 m
Downstream piezometric head = 2 m
Solution domain = 1500 m × 300 m
Discretization steps in x- and y-directions = 50 m and 10 m, respectively
Length and location of the sheet pile = 150 m deep at half distance of the domain (Figure 7.3)

The seepage flow under the dam can be estimated from the gradients (differences of the values of the flow potential) along the vertical line under the dam, extending from the dam or the tip of the sheet pile to the impermeable bed.

From the simulation results it can be seen that the computational error decreases very fast within the first 10 iteration cycles and then decreases asymptotically (Figure 7.4). However, in order for the phenomenon to reach steady-state conditions, the number of iterations was set equal to 1500. The reason for this large number of iterations is that since the initial conditions in the interior of the domain are set to zero potential head, during the first stages of the computation water flows from both the upstream and downstream boundary towards the

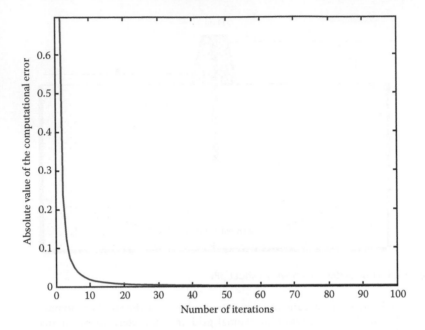

Figure 7.4 Number of iterations for convergence.

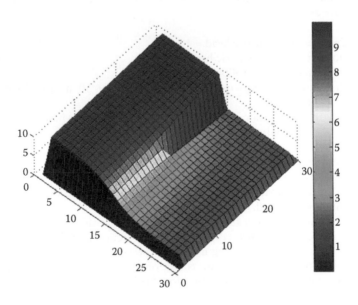

Figure 7.5 Three-dimensional presentation of the piezometric head distribution.

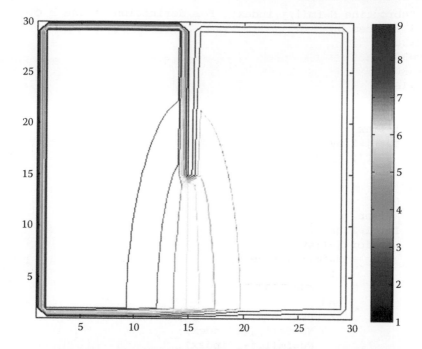

Figure 7.6 Two-dimensional presentation of the piezometric head distribution.

aquifer. That delays the establishment of the equilibrium conditions where water enters the aquifer from the ground upstream of the dam and exits from the ground downstream of the dam.

The distribution of the velocity potential under steady-state conditions is shown in three-dimensional rendering (Figure 7.5) and in side view (Figure 7.6). From Figures 7.5 and 7.6 it can be seen how the piezometric head reduces from its maximum value of 10 m at the left side to its minimum value of 2 m at the right side.

Computer code 7.1

```
% Example 7.1 Seepage Flow Under a Dam with Sheetpile
% f = Velocity potential [m^2/s];
% N = Numerical index defining the boundary conditions;
% Dx = Spatial step in the x-direction[m];
% Dy = Spatial step in the y-direction [m];
% nx = Number of spatial steps in the x-direction;
% ny = Number of spatial steps in the y-direction;
% nn = Number of iterations;
clc; clear all; close all;
load indices.txt
index=indices;
```

```
% Subroutine N=fm(i,j,index): For definition of the b.c.
  % indices nn;
nx=30;
ny=30;
Dx=50;
Dy=10;
nn=1500;
% Initialization of the variable;
for i=1:nx
    for j=1:ny
        f(i,j)=0;
    end
end
chk=0;
for k=1:nn
    for i=1:nx
        for j=1:ny
            N=fm(i,j,index);
            if N>0
                chk=1000;
            end
            if chk~=1000
                No=fm(i,j+1,index);
                Nu=fm(i,j-1,index);
                Nr=fm(i+1,j,index);
                Nl=fm(i-1,j,index);
                fo=f(i,j+1);
                fu=f(i,j-1);
                fr=f(i+1,j);
                fl=f(i-1,j);
                % Impervious boundary;
                if No==1
                    fo=f(i,j);
                end
                if Nu==1
                    fu=f(i,j);
                end
                if Nr==1
                    fr=f(i,j);
                end
                if Nl==1
                    fl=f(i,j);
                end
                % Constant head boundary;
                if No>1
                    fo=No;
                end
                if Nu>1
                    fu=Nu;
                end
```

```
                        if Nr>1
                            fr=Nr;
                        end
                        if Nl>1
                            fl=Nl;
                        end
                        temp=f(i,j);
                        f(i,j)=((fr+fl)/Dx^2+(fo+fu)/Dy^2)/(2/Dx^2+2/
                        Dy^2);
                        dif=abs(temp-f(i,j));
                end
                chk=0;
            end
        end
        diff(k)=dif;
        kk=k;
        fnew=f;
end
i=1:nx;
j=1:ny;
fpp=fnew(i,j);
plot(1:nn,diff,'b','Linewidth',1.5)
xlabel('Number of iterations');
ylabel('Absolute value of the computational error')
axis([0,100,0,0.7]);
figure
surf(i,j,fpp')
figure
contour(i,j,fpp')
```

PROBLEM 7.1

By making the suggested changes in the program while keeping the rest of the data unchanged, run the simulations, analyse the data and comment on the results.

1. Set $\Delta x = 100$ m and change the $\Delta x/\Delta y$ ratio to 1, 2 and 4.
2. Holding the upstream head equal to 10 m, change the tail water to 0.1, 4.0 and 8.0 m.
3. Change the pile sheet depth to 10, 50 and 250 m.
4. Assume that the pile sheet has failed at a depth between 50 and 75 m and allows groundwater flow through the opening. How will the seepage pattern change?
5. Considering that $\dfrac{\partial \varphi}{\partial x} = -\dfrac{\partial \psi}{\partial y}$ and $\dfrac{\partial \phi}{\partial y} = \dfrac{\partial \psi}{\partial x}$, where ψ is the streamline function, modify the code so that it estimates and plots the streamline function.

7.2.2 Horizontal two-dimensional groundwater flows

In the case of horizontally two-dimensional flows, the equations for confined (Equation 7.12) and unconfined flows (Equation 7.22) can be written in a unified format as

$$S_x \frac{\partial h}{\partial t} = \frac{\partial}{\partial x}\left(K_x H \frac{\partial h}{\partial x}\right) + \frac{\partial}{\partial y}\left(K_y H \frac{\partial h}{\partial y}\right) \pm q \tag{7.29}$$

where the 'storativity' S_x is either S (confined) or S_{ya} (unconfined), H is $b - \zeta_b$ (confined) or $h - \zeta_b$ (unconfined) (ζ_b is the bed elevation from some reference datum), and $q = \dfrac{Q}{\Delta x \Delta y}$ is the point discharge Q, averaged over a cell area, of a production (–) or recharging (+) well. Equation 7.29 is a linear parabolic equation in the case of a confined aquifer and a nonlinear parabolic equation in the case of an unconfined aquifer. The initial conditions were assumed either as a pre-existing flow distribution, or a horizontal free surface (unconfined aquifers) or piezometric surface (confined aquifers), implying no initial flow. The most common boundary conditions are either impermeable lateral boundaries $\dfrac{\partial h}{\partial n} = 0$ or constant head boundaries (h = constant).

The numerical solution can be done by an explicit central second-order finite differences scheme, or by the implicit Crank Nicolson scheme, combined with the ADI (alternative directions implicit) technique. The ADI technique consists of solving implicitly along one direction and explicitly along the other for one time step, and vice versa for the next time step. This approach for an N × N grid leads to the solution of N number of systems of N unknowns each, instead of the solution of a system in N^2 unknowns.

Using the Dupuit approximation, the water fluxes in the x- and y-directions for an isotropic aquifer can be defined as

$$q_x = -KH \frac{\partial h}{\partial x}, \quad q_y = -KH \frac{\partial h}{\partial y} \tag{7.30}$$

Thus, Equation 7.29 can be modified as follows:

$$S_x \frac{\partial h}{\partial t} = -\frac{\partial q_x}{\partial x} - \frac{\partial q_y}{\partial y} \pm q \tag{7.31}$$

The discretization of Equation 7.31 is accomplished by using a staggered two-dimensional explicit scheme where the fluxes are estimated on grid cell sides and the hydraulic head in the centre of the cell, leading to

$$S_x \frac{h_{i,j}^{n+1} - h_{i,j}^n}{\Delta t} = -\frac{q_{x_{i+1,j}}^n - q_{x_{i,j}}^n}{\Delta x} - \frac{q_{y_{i,j+1}}^n - q_{y_{i,j}}^n}{\Delta y} \pm q \tag{7.32}$$

where

$$q_{x_{i,j}} = -K_{i,j}(H_{i,j} + H_{i-1,j}) \frac{h_{i,j} - h_{i-1,j}}{2\Delta x} \tag{7.33}$$

$$q_{x_{i+1,j}} = -K_{i,j}(H_{i+1,j} + H_{i,j}) \frac{h_{i+1,j} - h_{i,j}}{2\Delta x} \tag{7.34}$$

$$q_{y_{i,j+1}} = -K_{i,j}(H_{i,j+1} + H_{i,j}) \frac{h_{i,j+1} - h_{i,j}}{2\Delta y} \tag{7.35}$$

$$q_{y_{i,j}} = -K_{i,j}(H_{i,j} + H_{i,j-1}) \frac{h_{i,j} - h_{i,j-1}}{2\Delta y} \tag{7.36}$$

The flow depth H is computed afterwards through the relation $H = h - \zeta_b$ from the computed h values and the known bed elevation ζ_b. The preceding numerical scheme needs to satisfy the stability criterion for parabolic equations:

$$\frac{K}{S_x} H_{max} < \frac{1}{2} \frac{(\Delta x)^2}{\Delta t} \tag{7.37}$$

Example 7.2

This exercise simulates a time-dependent, two-dimensional horizontal flow in an unconfined aquifer. The flow is induced by a constant rate pumping well located at the centre of the solution domain. The computer model is based on Equation 7.32 and Equations 7.33 to 7.36. The storativity S_x was taken as the porosity (voids ratio) n. All of the discretization nodes were identified with an index N. For internal points N = 0, for impervious boundaries N = 1, and for constant head boundary n is a real number different than 0 or 1 expressed in metres. Finally, the N corresponding to the location of the well is indicative of the well flow rate in cubic metres per hour (m³/hr). Other data used for the simulation are as follows:

Permeability coefficient = 0.5 m/hr
Porosity = 0.2
Area of the solution domain = 15,000 m × 15,000 m
Spatial steps equal in both x- and y-directions = 500 m
Time step = 24 hours

The simulation was conducted for a period of 7200 time steps (20 years). The pumping rate of the well is constant and equal to 400 m³/hr.

From the simulation results it is evident that the drop of the water elevation at the well site decreases exponentially in time (Figure 7.7). The simulation continues until there is a balance between the discharging well and the water entering the domain from the constant head boundary. In some cases, however, the simulation will continue until the water elevation in the well becomes zero. After that point the simulation ends.

Regarding the drawdown effects, the cone of depression is clearly shown in Figures 7.8 and 7.9. The effects of the impervious and the constant head boundary (left side in Figure 7.8) can be seen from the asymmetry developed in the shape of the water table surface.

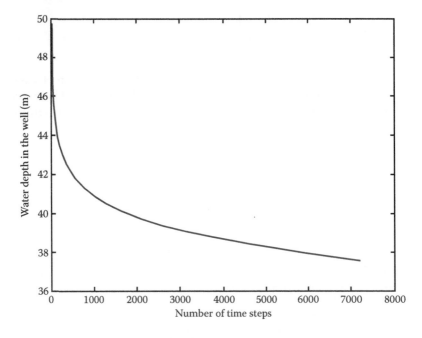

Figure 7.7 Water depth elevation changes over the time.

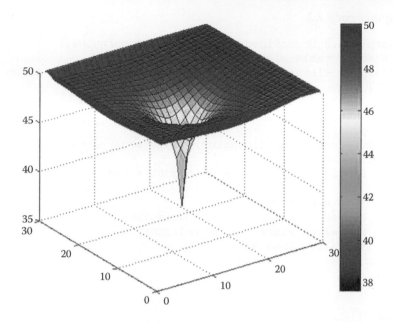

Figure 7.8 Three-dimensional view of the water table.

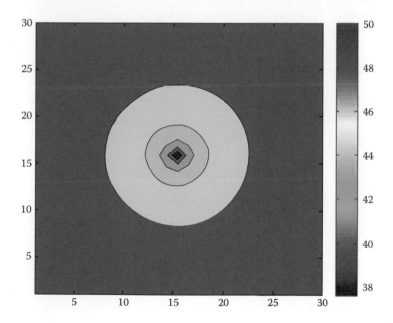

Figure 7.9 Plan view of the water table around the well.

Computer code 7.2

```
% Example 7.2 Unconfined Aquifer with Pumping Well
% h = Water table elevation [m];
% zb(j,j) = Elevation of the bed from the reference datum
  %[m];
% h(i,j) = Water surface elevation [m];
% K = Permeability coefficient [m/hr];
% p = Porosity;
% n = Node parameter (interior, imperveous, constant head,
  % well site);
% Dd = Spatial step (the same in both x and y directions
  %[m];
% Dt = Time step [hr];
% nx = Number of steps in the x-direction;
% ny = Number of steps in the y-direction;
clc; clear all; close all;
% Input data;
Dd=500;
Dt=24;
nx=30;
ny=30;
nt=7200;
K=0.5;
p=0.2;
rr=Dt/Dd^2/2*K/p;
% Initialization of variables;
for i=1:nx
    for j=1:ny
        h(i,j)=50;
        zb(i,j)=0;
        % n=0 indicates the interior nodes to be
        % calculated;
        n(i,j)=0;
    end
end
nxm=nx/2;
nym=ny/2;
% Definition of the boundary conditions;
for j=1:ny
    % n=1 indicates impervious boundaries;
    n(1,j)=1;
    n(nx,j)=1;
end
for i=1:nx
    n(i,1)=1;
    % n different than 0 and 1 indicates Piezometric head in
      % meters;
    n(i,ny)=50;
end
```

```
% Definition of the well location and head;
% n at the well site indicates flow in m^3/hr;
n(nxm,nym)=-400;
for i=1:nx
    for j=1:ny
        hn(i,j)=h(i,j);
    end
end
chk=0;
% Main program;
for k=1:nt
    for j=1:ny
        for i=1:nx
            if n(i,j)>0
                % chk is a number used for computational
                    % purposes;
                chk=1000;
            end
            if chk ~=1000
                fo=h(i,j+1);
                fu=h(i,j-1);
                fr=h(i+1,j);
                fl=h(i-1,j);
                % Impervious boundary conditions;
                if n(i,j+1)==1
                    fo=h(i,j);
                end
                if n(i,j-1)==1
                    fu=h(i,j);
                end
                if n(i+1,j)==1
                    fr=h(i,j);
                end
                if n(i-1,j)==1
                    fl=h(i,j);
                end
                % Constant head boundary conditions;
                if n(i,j+1)>1
                    fo=n(i,j+1);
                end
                if n(i,j-1)>1
                    fu=n(i,j-1);
                end
                if n(i+1,j)>1
                    fr=n(i+1,j);
                end
                if n(i-1,j)>1
                    fl=n(i-1,j);
                end
                dr=fr-zb(i+1,j);
```

```
                        dl=fl-zb(i-1,j);
                        do=fo-zb(i,j+1);
                        du=fu-zb(i,j-1);
                        dc=h(i,j)-zb(i,j);
                        fc=h(i,j);
                        Qwell=0;
                        if n(i,j)<0
                            Qwell=n(i,j)/Dd^2/p;
                        end
                        hn(i,j)=h(i,j)+rr*((dr+dc)*
                        (fr-fc)-(dc+dl)*(fc-fl)+(do+dc)*(fo-fc)-
                        (dc+du)*(fc-fu));
                        hn(i,j)=hn(i,j)+Dt*Qwell;
                    end
                    chk=0;
                end
            end
            for i=1:nx
                for j=1:ny
                    h(i,j)=hn(i,j);
                end
            end
            test=k
            WellDepth(k)=h(nxm,nym);
        end
        m=1:nt;
        WD=WellDepth(m);
        figure, plot(m,WD,'b','Linewidth',1.5);
        xlabel('Number of time steps');
        ylabel('Water depth in the well [m]');
        i=1:nx;
        j=1:ny;
        fpp=hn(i,j);
        figure, surf(i,j,fpp)
        figure
        contourf(i,j,fpp)
```

PROBLEM 7.2

Modify the application by making the following suggested changes while
keeping the rest of the data unchanged. Run the simulations, and comment
and explain the results.

1. Change the hydraulic conductivity to 0.05, 0.1, 0.25 and 0.75 m/hr.
2. Change the porosity to 0.1, 0.3 and 0.4.
3. Along with the pumping well, add a 100 m³/hr recharging well at
   node (i = 25, j = 25).

4. Change the (i, 0) boundary to constant head of 45 m.
5. Using the Darcy equation, modify the code of the program to estimate and plot the flow velocity field towards the well.

7.2.3 Saltwater intrusion in coastal aquifers

An important problem of coastal aquifers is the intrusion of salt water
(saline wedge) into the groundwater, due to any prolonged operation of
discharging (production) wells. This is a very common phenomenon during the summer months in Mediterranean coastal aquifers, when both the
increase of the population (due to summer vacations) and the use of irriga-
tion water reach an annual peak. By the end of the summer period, water
of increased salinity enters the production wells, to the disadvantage of the
local residents and the degradation of agricultural yield.

Based on the Dupuit assumption of nearly horizontal flow in an aquifer
with horizontal dimensions much larger than the vertical ones, a simulation model can be derived utilizing hydrostatic pressure distribution and
almost uniform seepage velocities throughout the porous medium (Bear
et al. 1999). In addition, a sharp interface is assumed between the overlay-
ing fresh water (of density ρ_o) and the underlying salt water (of density ρ_u).
For an unconfined aquifer, the situation is illustrated in Figure 7.10.

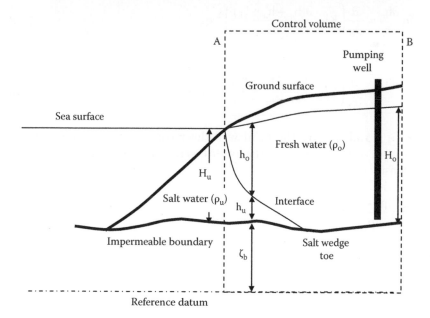

Figure 7.10 Saline wedge intrusion in a coastal aquifer.

The modelled area extends from the coast to an upstream boundary where the water table in the ground is assumed either constant (adequate recharge from water infiltrating in the ground upstream of the hydrological basin) or of predefined slope (limited recharge from upstream). On the coastal boundary the salt water depth is constant and the fresh water layer becomes negligible. The two immiscible layers are characterized by their thickness (h_o and h_u) varying with time along the horizontal spatial directions ($h_o = h_o(x,y,t)$) and their constant densities ρ_o and ρ_u.

According to the Dupuit assumption, the specific discharges in the two layers (of the almost horizontally flowing water) are related to the pressure gradient along the horizontal dimensions. Gravity is also taken into consideration since the interface between the two layers and the impermeable bottom of the aquifer are inclined ('mild' slopes are assumed) (Arvanitidou, Katsifarakis and Koutitas 2012).

7.2.3.1 One-dimensional equations for saltwater intrusion

Application of the continuity equation to a control water column of base Δx and Δy in both the upper and the lower layers lead to two PDEs relating the time variation of the h_o and h_u with their spatial gradients, the aquifer permeability $K(x,y)$, and the soil porosity n (the void ratio).

In the simplest form, the model, in one spatial direction, reads

$$n\frac{\partial h_o}{\partial t} = \frac{\partial}{\partial x}\left(Kh_o\frac{\partial(h_o+h_u+\zeta_b)}{\partial x}\right) - q_{wo} \tag{7.38}$$

$$n\frac{\partial h_u}{\partial t} = \frac{\partial}{\partial x}\left(Kh_u\frac{\partial(h_o+h_u+\zeta_b)}{\partial x} - Kh_u\delta\frac{\partial h_o}{\partial x}\right) - q_{wu} \tag{7.39}$$

where q_{wo} and q_{wu} are the local well discharges pumping or recharging water from or to the upper or the lower layer, respectively, and δ is the relative density difference defined as

$$\delta = \frac{\rho_u - \rho_o}{\rho_u} \tag{7.40}$$

The boundary conditions at sides A and B of the control volume (Figure 7.10) are defined as

Left side boundary A: $h_{oA} = 0$ and $h_{uA} = H_u$

Right side boundary B: $h_{oB} = H_o$ or $\left(\dfrac{\partial h_{oB}}{\partial x} = \text{defined}\right)$ and $h_{uB} = 0$

The last boundary condition implies that the toe of the saline wedge, propagating under the fresh water layer, does not reach the end of the modelled control volume area.

7.2.3.2 Computational scheme of the governing equations

The form of the governing Equations 7.38 and 7.39 allows the decoupling of the solution for the two basic variables h_o and h_u and the utilization of the time derivatives of h_o and h_u appearing on the left-hand side of the two equations for the explicit propagation of the solution from a time level (n) to the next (n + 1). After discretization of the domain, the numerical solution is accomplished on a staggered grid, where the seepage specific discharges Q, are computed on the sides of a grid cell and the layers thickness h_o, h_u at the cell's centre. The numerical scheme first estimates the fluxes Q_o and Q_u from known data at time n, and then computes the layers thickness h_o and h_u at the new time step n + 1. The explicit FTCS (forward in time, central in space) finite differences scheme is given by Equations 7.41 through 7.46:

$$Q_{o,i} = -K \frac{\left(h_{o,i}^n + h_{u,i}^n + \zeta_{b,i} - h_{o,i-1}^n - h_{u,i-1}^n - \zeta_{b,i-1}\right)}{\Delta x} \cdot \frac{\left(h_{o,i}^n + h_{o,i-1}^n\right)}{2} \qquad (7.41)$$

$$Q_{o,i+1} = -K \frac{\left(h_{o,i+1}^n + h_{u,i+1}^n + \zeta_{b,i+1} - h_{o,i}^n - h_{u,i}^n - \zeta_{b,i}\right)}{\Delta x} \cdot \frac{\left(h_{o,i}^n + h_{o,i+1}^n\right)}{2} \qquad (7.42)$$

$$h_{o,i}^{n+1} = -h_{o,i}^n - \frac{1}{n} \frac{\Delta t}{\Delta x}(Q_{o,i+1} - Q_{o,i} + q_{wo}) \qquad (7.43)$$

$$Q_{u,i} = -K \frac{\left(h_{o,i}^n + h_{u,i}^n + \zeta_{b,i} - h_{o,i-1}^n - h_{u,i-1}^n - \zeta_{b,i-1} - \delta\left(h_{o,i}^n - h_{o,i-1}^n\right)\right)}{\Delta x} \cdot \frac{\left(h_{u,i}^n + h_{u,i-1}^n\right)}{2} \qquad (7.44)$$

$$Q_{u,i+1} = -K \frac{\left(h_{o,i+1}^n + h_{u,i+1}^n + \zeta_{b,i+1} - h_{o,i}^n - h_{u,i}^n - \zeta_{b,i} - \delta\left(h_{o,i+1}^n - h_{o,i}^n\right)\right)}{\Delta x}$$
$$\cdot \frac{\left(h_{u,i}^n + h_{u,i+1}^n\right)}{2} \qquad (7.45)$$

$$h_{u,i}^{n+1} = -h_{u,i}^n - \frac{1}{n} \frac{\Delta t}{\Delta x}(Q_{u,i+1} - Q_{u,i} + q_{wu}) \qquad (7.46)$$

where q_{wo} and q_{wu} denote the discharge from a pumping well from the fresh water (o) and the saline water (u) layers.

Example 7.3

This exercise simulates the formation of a salt wedge in a one-dimensional coastal unconfined aquifer under the effects of a production well. The base of the aquifer is horizontal and impermeable, and fresh water is supplied from the upstream part of the hydrological basin. The aim is to evaluate the effects of the production well on the saline wedge interface. This information can assist in water resource management decisions to avoid saltwater intrusion. The data used for the simulation are as follows:

Hydraulic conductivity = 500 m/day
Soil porosity = 0.4
Upper layer pump discharge = 3 m³/m/day
Lower layer pump discharge = 0 m³/m/day
Freshwater thickness at the inland boundary = 20.8 m
Saltwater thickness at the coastal boundary = 20 m
Relative density difference = 0.02
Longitudinal distance of the solution domain = 10,000 m
Distance of the upper layer pumping well from the coast = 9000 m
Distance of the lower layer pumping well from the coast = 0 m

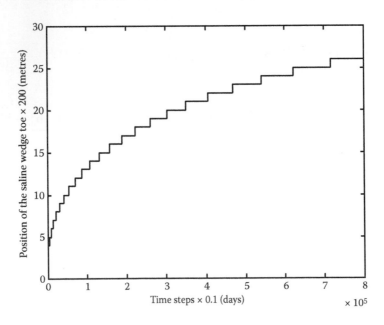

Figure 7.11 Advancement of the saline wedge toe.

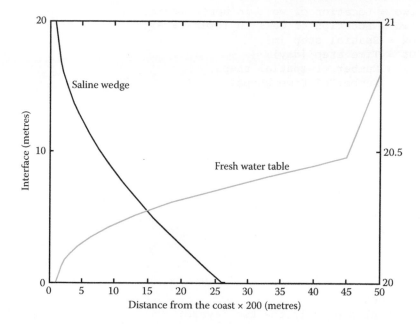

Figure 7.12 Saline wedge and groundwater table affected by the pumping well.

The solution domain was discretized into spatial steps $\Delta x = 200$ m and the time step was taken as $\Delta t = 0.1$ days. The initial conditions were set so that at the beginning the saline water was confined only to the coastal boundary at a depth equal to the entire water column. Then the density difference generated a gravity current creating the advancement of the salt wedge below the fresh water lens. The simulation was conducted for a period of 800,000 time steps, until quasi-steady-state conditions were established. The saline wedge propagated to a distance of about 5400 metres from the coast. The advancement of the saline wedge toe in the inland direction is illustrated in Figure 7.11. The quasi-steady-state profiles of the saline wedge and the water table are shown in Figure 7.12. The effects of the pumping well located at node 45 (9000 m) is documented in Figure 7.12.

Computer code 7.3

```
% Example 7.3 Salt Water Intrusion Due to Pumping Well
% Ho = Initial depth of the upper layer [m];
% Hu = Initial depth of the lower layer [m];
% qo = Point discharge of the upper layer [m^3/m/day];
% qu = Point discharge of the lower layer [m^3/m/day];
% p = Porosity;
% K = Permeability coefficient [m/day];
% rdf = Relative density difference;
```

```
% iwo = Location of well in upper layer;
% iwu = Location of well in lower layer;
% Dx = Spatial step [m];
% Dt = Time step [day];
% nx = Number of spatial steps;
% nt = Number of time steps;
clc; clear all; close all;
% Input data;
Ho=20.8;
Hu=20;
qo=-3;
qu=0;
p=0.4;
K=500;
rdf=0.02;
iwo=45;
iwu=0;
Dx=200;
Dt=0.1;
nx=50;
nt=800000;
% Initial depths in the two layers;
for i=2:nx-1
    ho(i)=Ho;
    hon(i)=ho(i);
    hu(i)=0;
    hun(i)=hu(i);
end
ho(nx)=Ho;
hon(nx)=Ho;
hu(1)=Hu;
hun(1)=Hu;
ho(1)=0;
hon(1)=ho(1);
hu(nx)=0;
hun(nx)=hu(nx);
for k=1:nt
    % Calculation of new depth values of the upper layer;
    for i=2:nx-1
        Qwo=0;
        if i==iwo
            Qwo=qo;
        end
        qr=-K*(ho(i+1)+hu(i+1)-ho(i)-hu(i))/
        Dx*(ho(i)+ho(i+1))/2;
        ql=-K*(ho(i)+hu(i)-ho(i-1)-hu(i-1))/
        Dx*(ho(i)+ho(i-1))/2;
        hon(i)=ho(i)-Dt*(qr-ql-Qwo)/Dx/p;
    end
```

```
    % Calculation of new depth values of the lower layer;
    for i=2:nx-1
        Qwu=0;
        if i==iwu
            Qwu=qu;
        end
        qr=-K*(ho(i+1)+hu(i+1)-ho(i)-hu(i)-rdf*(ho(i+1)-
        ho(i)))/Dx*(hu(i)+hu(i+1))/2;
        ql=-K*(ho(i)+hu(i)-ho(i-1)-hu(i-1)-rdf*(ho(i)-
        ho(i-1)))/Dx*(hu(i)+hu(i-1))/2;
        hun(i)=hu(i)-Dt*(qr-ql-Qwu)/Dx/p;
        if hun(i)<0
            hun(i)=0;
        end
    end
    % Boundary conditions;
    hun(nx)=hun(nx-1);
    % Updating depth values;
    for i=1:nx
        ho(i)=hon(i);
        hu(i)=hun(i);
    end
    chk=0;
    xtoe=0;
    % Estimation of saline wedge toe location;
    for i=1:nx
        if hu(i)<0.1
            xtoe=i;
            break
        end
    end
    toe(k)=xtoe;
    index=k
end
i=1:nx;
GWt=ho(i);
SWe=hu(i);
ground = GWt+SWe;
plot(1:k,toe,'b','Linewidth',1.5)
xlabel('Time steps x 0.1 [days]')
ylabel('Position of the saline wedge toe x 200 [meters]')
figure
plotyy(1:nx,SWe,1:nx,ground)
xlabel('Distance from the coast x 200 [meters]')
ylabel('Interface [meters]')
text(3,15,'Saline wedge');
text(25,8.8,'Fresh water table');
```

PROBLEM 7.3

Modify the application by making the following suggested changes while keeping the rest of the data unchanged. Conduct the computer simulations, then comment and explain the results.

1. Increase the pumping rate to 1.0, 6.0 and 9.0 m^3/m/day.
2. Move the location of the well to a distance of 5 km from the shoreline.
3. Change the hydraulic conductivity to 200 and 1000 m/day.
4. Change the relative density difference to 0.01, 0.035 and 0.05.
5. At a distance of 5 km from the shoreline add a recharging well of 3.0 m^3/m/day.

Chapter 8

Contaminant and sediment transport by advection and diffusion

8.1 INTRODUCTION TO MATHEMATICAL MODELLING OF TRANSPORT PROCESSES

8.1.1 General concepts

In nature, matter is transported exclusively by means of a moving fluid. Thus, in geophysical systems pollutants are transported either by moving water or moving air. Without those fluids, any pollution source would remain local. The transport of matter is achieved in two ways:

1. In solution, when the matter is dissolved in the fluid. In this case, the molecules of the matter are mixed with the molecules of the fluid and retained in that condition, due to electrochemical forces acting at molecular scale. Any difference of density between the matter and the carrying fluid is not important for concentrations less than 1% (weight or volume of dissolved matter per unit volume of the solution).

2. In suspension, when the matter is in particulate form. The particles, being much bigger than the molecules of the fluid, are suspended within the fluid mass. In this case the suspension process is controlled by the density of the matter relative to the suspending fluid, and the driving force is the gravity rather than the electrochemical molecular forces. Suspended matter can either be subject to settling (sedimentation), for particles with density larger than that of water, or floating (flotation due to buoyancy), for particles with density smaller than the water density.

In both cases, the amount of the matter in the fluid is expressed by its concentration, that is, the ratio of the volume (or mass) of the matter over the volume (or mass) of the mixture (in dissolved or suspended state). The units are usually milligrams per litre (mg/L) or a dimensionless number. In general, the concentration, c, is a function of space and time, c(x,y,z,t).

For solutions the density difference between the dissolved matter and the fluid is not important, except in case of very high concentrations. The behaviour of suspensions is characterized by two mechanisms: (1) the inertial transient behaviour of the suspended matter, before it obtains a velocity equal to that of the carrying fluid, and (2) the tendency of the suspended particles to move vertically upwards (float) or downwards (settle). Floating particles may arrive to the surface and stay there, while settling particles may reach the bed and either stay there or be re-suspended. The final deposition or re-suspension depends on the particle's geometry and density, the properties of the ambient fluid and the flow field. The intensity of the flow field, in terms of local turbulence, is quantified by the value of the local eddy viscosity/diffusivity coefficient. A differentiation between coarse sand and silt/clay (cohesive) particles has to be made since cohesive particles are subject to surface electrochemical forces (Scarlatos 2002).

8.1.2 Mathematical formulation

The transport of matter is caused by advection and diffusion processes. Advection, after some short 'transient period', results in the matter moving with a velocity equal to the fluid velocity. Diffusive transport, according to the Fick's law, is always in the direction of the reducing concentrations, controlled by the molecular diffusion or (in the case of turbulent flows) by the eddy diffusion. Thus the transport components are quantified by the advective flux (f_a) and the diffusive flux (f_d). In the x-axis direction, the advective flux is described as

$$f_{ax} = uc \tag{8.1}$$

where c is the concentration of matter and u is the local fluid velocity in the x-axis direction. The diffusive flux, also in the x-axis direction, is given by the relation

$$f_{dx} = -\nu \frac{\partial c}{\partial x} \tag{8.2}$$

where ν is the molecular diffusion coefficient (kinematic viscosity of the fluid). In turbulent flow regimes, the random fluctuations of the advective velocity components (u', v', w') cause a transport effect similar to that of molecular diffusion but much more intensive. In that case, the molecular diffusion becomes negligible as compared to the eddy (or turbulent) diffusion. This type of diffusion is parameterized by an eddy diffusion coefficient (D_t), directly related to the eddy viscosity (ν_t). Then the turbulent diffusive flux, f_{tx}, also described by Fick's law, becomes

$$f_{tx} = -(v + D_t)\frac{\partial c}{\partial x} \approx -D_t \frac{\partial c}{\partial x} \tag{8.3}$$

The turbulent diffusion coefficient depends on the flow characteristics (the mean velocities gradients and their turbulent fluctuations). In general, determination of the function $D_t(x,y,z,t)$ is a very complex process following the difficulties involved when solving for the turbulent mean velocities. However, the process can be simplified by assuming two separate eddy diffusion parameters, D_h and D_v, where D_h refers to the horizontal mixing and D_v to the vertical mixing. The difference between the two parameters stems from the fact that in most geophysical horizontal flows, water depths are much smaller as compared to the horizontal dimensions of the flow domain. Therefore, the development of turbulent eddies have different geometric and energy content, and characteristics between the vertical and the horizontal dimensions.

Development of a mass transport mathematical model for the estimation of $c(x,y,z,t)$ can be accomplished by using the following basic considerations:

1. The transported substance does not interfere with the pre-existing hydrodynamic conditions. This is true for concentrations less than 0.1 ppt.
2. The transported substance may be conservative or non-conservative, that is, its total mass may remain constant during the transport, or it may decay due to mechanical, chemical or biological processes which evolve with the transport. For instance, biological decomposition is usually described by an exponential decay law ($c = c_o \exp(-\lambda t)$, where λ is the bio-decomposition rate (with dimensions, s^{-1}).

For a one-dimensional system, application of the mass conservation principle on a linear control space (Figure 8.1), and utilization of Equations 8.1 and 8.3 leads to the following mass transport equation:

$$\frac{\partial c}{\partial t} + \frac{\partial(uc)}{\partial x} = \frac{\partial}{\partial x}\left(D_t \frac{\partial c}{\partial x}\right) \tag{8.4}$$

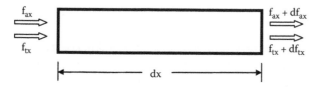

Figure 8.1 One-dimensional control volume for mass flux.

For a conservative substance in suspension settling with a velocity w_f, Equation 8.4 can be written in a three-dimensional form as

$$\frac{\partial c}{\partial t} + \frac{\partial (uc)}{\partial x} + \frac{\partial (vc)}{\partial y} + (w - w_f)\frac{\partial c}{\partial z} = \frac{\partial}{\partial x}\left(D_h \frac{\partial c}{\partial x}\right) + \frac{\partial}{\partial y}\left(D_h \frac{\partial c}{\partial y}\right)$$

$$+ \frac{\partial}{\partial z}\left(D_v \frac{\partial c}{\partial z}\right) + Sources - Sinks \qquad (8.5)$$

For completeness of the mathematical model, the free surface and bed boundary conditions are described, respectively, as

$$cw_f = -D_v \frac{\partial c}{\partial z}\Big|_{z=0} \qquad (8.6)$$

$$c(1 - r)w_f = -D_v \frac{\partial c}{\partial z}\Big|_{z=-h} \qquad (8.7)$$

where r accounts for the re-suspension. For dissolved matter, the preceding equations hold for $w_f = 0$.

By defining the depth-integrated velocities and concentration as

$$U = \frac{1}{h} \int_{z=-h}^{z=0} udz \text{ and } V = \frac{1}{h} \int_{z=-h}^{z=0} vdz \qquad (8.8)$$

$$C = \frac{1}{h} \int_{z=-h}^{z=0} cdz \qquad (8.9)$$

the model in terms of the depth-averaged concentration $C(x,y,t)$ is modified to

$$\frac{\partial C}{\partial t} + \frac{\partial (UC)}{\partial x} + \frac{\partial (VC)}{\partial y} = \frac{1}{h}\frac{\partial}{\partial x}\left(hK_h \frac{\partial C}{\partial x}\right) + \frac{1}{h}\frac{\partial}{\partial y}\left(hK_h \frac{\partial C}{\partial y}\right) - \left(\lambda + \frac{rw_f}{h}\right)C$$

$$(8.10)$$

The newly appearing parameter is the dispersion coefficient K_h ($> D_h, D_v$) resulting from the non-uniform velocity distribution along the depth. The settling velocity is taken as negative. Both forms of the transport model (Equation 8.5 and Equation 8.10) indicate that the problem involves the solution of a mixed-type equation hyperbolic–parabolic, that is, describing a concurrent advection and diffusion process. The relative importance of the two processes is quantified through the Peclet number, $P_e = \dfrac{UL}{D}$, where L is a characteristic spatial scale of the phenomenon. $P_e < 1$ indicates that the diffusion prevails, while $P_e > 1$ indicates that the advection is the predominant process.

8.2 NUMERICAL SOLUTIONS OF THE TRANSPORT MODEL

The effects of numerical diffusion and dispersion of hyperbolic equations on the accuracy of the computational results have been discussed in Chapter 3. However, the numerical solution of the general transport model, governed by a mixed hyperbolic–parabolic equation, poses a unique computational challenge. For instance, if a 'diffusive' numerical scheme is selected, depending on the Peclet number, the numerical diffusion may distort the effects of the naturally occurring diffusion.

An explicit finite differences scheme can be obtained by using, for the advection term, forward differences for the time derivative, and backward differences for the space derivative (FTBS, forward in time, backward in space); and, for the diffusion term, a second-order centred differences scheme:

$$\frac{C_i^{n+1} - C_i^n}{\Delta t} + U_i \frac{C_i^n - C_{i-1}^n}{\Delta x} = D \frac{C_{i+1}^n - 2C_i^n + C_{i-1}^n}{(\Delta x)^2} \tag{8.11}$$

or

$$C_i^{n+1} = C_i^n - U_i \frac{\Delta t}{\Delta x}\left(C_i^n - C_{i-1}^n\right) + D \frac{\Delta t}{(\Delta x)^2}\left(C_{i+1}^n - 2C_i^n + C_{i-1}^n\right) \tag{8.12}$$

It should be noted that for negative U, the second term in Equation 8.11 becomes $U_{i+1} \dfrac{C_{i+1}^n - C_i^n}{\Delta x}$ and changes accordingly in Equation 8.12. The preceding numerical scheme is stable but diffusive. For small Peclet numbers the real diffusion is high, and the numerical diffusion becomes insignificant. On the contrary, for Pe > 1, selection of a proper numerical scheme is very important, since the numerical diffusion may be larger than the physical one. In those cases it is recommended to apply either the Fromm scheme or total variation diminishing (TVD) scheme (see Chapter 3). More specifically, in the TVD scheme the backward diffusive difference used for the advection is combined with additive diffusive terms involving a negative 'diffusion' coefficient. Following the same procedure as in Chapter 3, Equation 8.12 can be re-written in the TVD scheme as

$$C_i^{n+1} = C_i^n - \omega\left(C_i^n - C_{i-1}^n\right) - \left(K_{i+\frac{1}{2}} - K_{i-\frac{1}{2}}\right) + D\frac{\Delta t}{(\Delta x)^2}\left(C_{i+1}^n - 2C_i^n + C_{i-1}^n\right) \quad (8.13)$$

$$\omega = \frac{U\Delta t}{\Delta x} \quad (8.14)$$

$$K_{i+\frac{1}{2}} = \frac{\omega}{2}(1-\omega)\left(C_{i+1}^n - C_i^n\right)\varphi(R_i) \quad (8.15)$$

$$K_{i-\frac{1}{2}} = \frac{\omega}{2}(1-\omega)\left(C_i^n - C_{i-1}^n\right)\varphi(R_i) \quad (8.16)$$

$$R_i = \frac{C_i^n - C_{i-1}^n}{C_{i+1}^n - C_i^n} \quad (8.17)$$

while $\varphi(R_i) = \min(2R_i, 2)$ for $R_i > 0$ and $\varphi(R_i) = 0$ for $R_i < 0$. The superiority of the TVD scheme as compared to the FTBS can be clearly identified for Peclet = 2.

Example 8.1

This application involves the simulation of the transport of a non-conservative substance in a horizontally two-dimension field with a marina breakwater structure. The governing equation (Equation 8.10) is discretized using a FTBS scheme similar to the one given by Equation 8.12. The bathymetry, flow field and the eddy diffusion coefficients are taken from the simulation results of Example 5.4 (Chapter 5). Thus, the same staggered discretization grid is applied, where the velocities U and V are defined at the sides of the grid cells, while the water surface elevation (h) is defined at the centre of the cells. The substance concentration is computed similarly with the free surface elevation at the centre of the grid cells. The advection term, after checking the sign of the sums $U_{i,j} + U_{i+1,j}$ (x-direction) and $V_{i,j} + V_{i,j+1}$ (y-direction) is expressed either by a backward or forward finite difference (see Equation 8.12). Other data provided are

Concentration (strength) of point source = 1 Kg/m³
Decay coefficient = 0 s⁻¹
Location of the point source: x = 90 m; y = 125 m

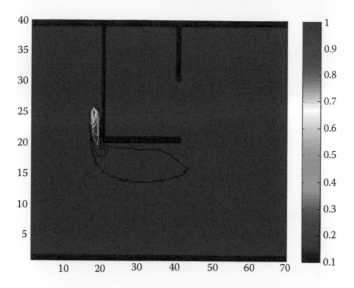

Figure 8.2 Advection and diffusion of a substance near a breakwater structure.

The time step was $\Delta t = 5$ s, and the simulation was conducted for a period of 720 time steps (3600 s). The simulation results are illustrated in Figure 8.2, where the effects of both advection and diffusion are clearly shown. The concentrations depicted in Figure 8.2 refer to the maximum concentration values estimated for each location over the entire simulation period.

Computer code 8.1

```
% Example 8.1 Contaminant Advective Diffusion Near Marina
Breakwater
% csource = Pollution concentration at the source [kg/m^3];
% dc = Decay coefficient [s^-1];
% imm = x-axis coordinate of the pollution source [number of
steps];
% jmm = y-axis coordinate of the pollution source [number of
steps];
% Dd = Spatial step (same in both directions) [m];
% Dt = Time step [s];
% nx = Number of computational steps in the x-direction;
% ny = Number of computational steps in the y-direction;
% nt = Number of time steps;
clc; clear all; close all;
% Input data;
csource=1;
dc=0.0;
imm=18;
```

```
jmm=25;
Dd=5;
Dt=5;
nx=70;
ny=40;
nt=720;
% Initialization of variables;
for i=1:nx
    for j=1:ny
        c(i,j)=0;
        cmax(i,j)=0;
        cn(i,j)=0;
    end
end
% Importing data for depths, velocities and diffusion
coefficients;
load Depths.txt
load CoastU.txt
load CoastV.txt
load EddyC.txt
fDep=Depths;
fCoU=CoastU;
fCoV=CoastV;
fEdd=EddyC;
% Subrouting for depths h=fm(i,j,fDep);
% Subroutine for velocity u=fm(i,j,fCoU);
% Subroutine for velocity v=fm(i,j,fCoV);
% Subroutine for diffusion coefficients ed=fm(i,j,fEdd);
% Main program;
for k=1:nt
    for j=2:ny-2
        for i=2:nx-2
            h=fm(i,j,fDep);
            ho=fm(i,j+1,fDep);
            hu=fm(i,j-1,fDep);
            hl=fm(i-1,j,fDep);
            hr=fm(i+1,j,fDep);
            u=fm(i,j,fCoU);
            ur=fm(i+1,j,fCoU);
            v=fm(i,j,fCoV);
            vo=fm(i,j+1,fCoV);
            ed=fm(i,j,fEdd);
            if h>0.1
                advx=0;
                if u+ur>0
                    advx=u*(c(i,j)-c(i-1,j))/Dd;
                else
                    advx=ur*(c(i+1,j)-c(i,j))/Dd;
                end
                advy=0;
```

```
        if v+vo>0
            advy=v*(c(i,j)-c(i,j-1))/Dd;
        else
            advy=vo*(c(i,j+1)-c(i,j))/Dd;
        end
        co=c(i,j+1);
        if ho<0.1
            co=c(i,j);
        end
        cu=c(i,j-1);
        if hu<0.1
            cu=c(i,j);
        end
        cr=c(i+1,j);
        if hr<0.1
            cr=c(i,j);
        end
        cl=c(i-1,j);
        if hl<0.1
            cl=c(i,j);
        end
        diff1=ed/h*((co-c(i,j))*(h+ho)-(c(i,j)-
        cu)*(h+hu))/2/Dd^2;
        diff2=ed/h*((cr-c(i,j))*(hr+h)-(c(i,j)-
        cl)*(h+hl))/2/Dd^2;
        cn(i,j)=c(i,j)+Dt*(-advx-advy+diff1+diff2-
        dc*c(i,j));
        end
    end
end
% Boundary conditions;
for i=2:nx-1
    cn(i,2)=cn(i,3);
    cn(i,ny-2)=cn(i,ny-3);
end
for j=2:ny-1
    cn(2,j)=2*cn(3,j)-cn(4,j);
    cn(nx-2,j)=2*cn(nx-3,j)-cn(nx-4,j);
end
% Point source pollution;
cn(imm,jmm)=csource;
% Updating the contaminant concentration values;
for j=1:ny
    for i=1:nx
        c(i,j)=cn(i,j);
        if cn(i,j)>cmax(i,j)
            cmax(i,j)=cn(i,j);
        end
    end
end
```

```
      index=k
end
[i,j]=meshgrid(1:1:nx,1:1:ny);
df1=fDep(:,3);
df2=reshape(df1,70,40);
df2=df2';
An=ones(size(df2));
idx=find(df2== 0);
An(idx)=0;
figure;
pcolor(An);
colormap(jet);
% colormap(gray(3))
shading flat;
hold on;
i=1:nx;
j=1:ny;
Cplot=cmax(i,j);
% hold on;
contour(i,j,Cplot')
% colorbar('vert')
```

PROBLEM 8.1

Solve the same problem by making the suggested modifications while keeping the rest of the data constant:

1. Place the contaminant point source at location x = 225 m, y = 190 m.
2. Remove the horizontal arm of the breakwater.
3. Replace the point source by a line source extending from point (50 m, 25 m) to point (50 m, 75 m).
4. Remove the left vertical section of the breakwater.
5. Change the decay coefficient to 0.01 s^{-1} (decay) and -0.01 s^{-1} (growth).

Run the simulations, compare the results and comment on the changes observed regarding the spreading of the contaminant plume.

Example 8.2

This application is based on the bathymetry and hydrodynamic data from Example 5.5 and investigates the advection and diffusion of a pollutant point source in the Thermaikos Gulf. The results are indicative of the contaminant transport under the forcing of a northwest wind of constant speed of 9.94 m/s. Other data used for the simulation are

Concentration (strength) of point source = 1 Kg/m^3
Decay coefficient = 10^{-5} s^{-1}
Location of the point source: x = 24 km; y = 24 km

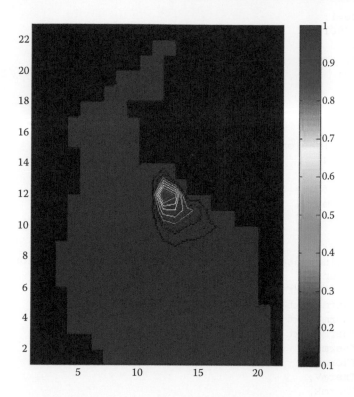

Figure 8.3 Contaminant advection and diffusion in the Thermaikos Gulf.

Using a time step Δt = 600 s (10 minutes), the simulation was conducted for 288 time steps (2 days), and the results are illustrated in Figure 8.3. From this figure the effects of advection and diffusion of the contaminant are clearly shown.

Computer code 8.2

```
% Example 8.2 Contaminant Advective Diffusion in Thermaikos
Gulf
% csource = Pollution concentration at the source [kg/m^3];
% dc = Decay coefficient [s^-1];
% imm = x-axis coordinate of the pollution source;
% jmm = y-axis coordinate of the pollution source;
% Dd = Spatial step (same in both directions) [m];
% Dt = Time step {s};
% nx = Number of computational steps in the x-direction;
% ny = Number of computational steps in the y-direction;
% nt = Number of time steps;
clc; clear all; close all;
% Input data;
```

```
csource=1;
dc=0.00001;
imm=12;
jmm=12;
Dd=2000;
Dt=600;
nx=22;
ny=23;
nt=288;
% Initialization of variables;
for i=1:nx
    for j=1:ny
        c(i,j)=0;
        cmax(i,j)=0;
        cn(i,j)=0;
    end
end
% Importing data for depths, velocities and diffusion
coefficients;
load ThermD.txt
load ThermU.txt
load ThermV.txt
load ThermE.txt
fDep=ThermD;
fCoU=ThermU;
fCoV=ThermV;
fEdd=ThermE;
% Subrouting for depths h=fm(i,j,fDep);
% Subroutine for velocity u=fm(i,j,fCoU);
% Subroutine for velocity v=fm(i,j,fCoV);
% Subroutine for diffusion coefficients ed=fm(i,j,fEdd);
% Main program;
for k=1:nt
    for j=2:ny-2
        for i=2:nx-2
            h=fm(i,j,fDep);
            ho=fm(i,j+1,fDep);
            hu=fm(i,j-1,fDep);
            hl=fm(i-1,j,fDep);
            hr=fm(i+1,j,fDep);
            u=fm(i,j,fCoU);
            ur=fm(i+1,j,fCoU);
            v=fm(i,j,fCoV);
            vo=fm(i,j+1,fCoV);
            ed=fm(i,j,fEdd);
            if h>0.1
                advx=0;
                if u+ur>0
                    advx=u*(c(i,j)-c(i-1,j))/Dd;
```

```
        else
            advx=ur*(c(i+1,j)-c(i,j))/Dd;
        end
        advy=0;
        if v+vo>0
            advy=v*(c(i,j)-c(i,j-1))/Dd;
        else
            advy=vo*(c(i,j+1)-c(i,j))/Dd;
        end
        co=c(i,j+1);
        if ho<0.1
            co=c(i,j);
        end
        cu=c(i,j-1);
        if hu<0.1
            cu=c(i,j);
        end
        cr=c(i+1,j);
        if hr<0.1
            cr=c(i,j);
        end
        cl=c(i-1,j);
        if hl<0.1
            cl=c(i,j);
        end
        diff1=ed/h*((co-c(i,j))*(h+ho)-(c(i,j)-
        cu)*(h+hu))/2/Dd^2;
        diff2=ed/h*((cr-c(i,j))*(hr+h)-(c(i,j)-
        cl)*(h+hl))/2/Dd^2;
        cn(i,j)=c(i,j)+Dt*(-advx-advy+diff1+diff2-
        dc*c(i,j));
        end
    end
end
% Boundary conditions;
for i=2:nx-1
    cn(i,2)=2*cn(i,3)-c(i,4);
end
%    for j=2:ny-1
%        cn(2,j)=2*cn(3,j)-cn(4,j);
%        cn(nx-2,j)=2*cn(nx-3,j)-cn(nx-4,j);
%    end
% Point source pollution;
cn(imm,jmm)=csource;
% Updating the contaminant concentration values;
for j=1:ny
    for i=1:nx
        c(i,j)=cn(i,j);
        if cn(i,j)>cmax(i,j)
```

```
                    cmax(i,j)=cn(i,j);
            end
        end
    end
    index=k
end
[i,j]=meshgrid(1:1:nx,1:1:ny);
df1=fDep(:,3);
df2=reshape(df1,22,23);
df2=df2';
An=ones(size(df2));
idx=find(df2== 0);
An(idx)=0;
figure;
pcolor(An); hold on;
colormap(jet)
shading flat
i=1:nx;
j=1:ny;
Cplot=cmax(i,j);
contour(i,j,Cplot')
% colorbar('vert')
```

PROBLEM 8.2
Solve the same problem by making the suggested modifications while keeping the rest of the data constant:

1. Place the contaminant point source at location x = 16 km, y = 34 km.
2. The source ceases to pollute after 1440 minutes.
3. Replace the decay coefficient with a growth coefficient of the same rate.
4. Place two additional pollutant sources at locations (28 km, 20 km) and (32 km, 16 km).
5. Change the time step to Δt = 120 min, 180 min and 240 min.

Run the simulations, compare the results and comment on the changes observed regarding the spreading of the contaminant plume.

8.3 LAGRANGIAN MODELLING OF MASS TRANSPORT

The hydrodynamic phenomena can be described by using either Lagrangian or Eulerian methodology. The Lagrangian description (moving coordinates) uses the concept of the 'system'. The system is always comprised of the same group of fluid particles and the Lagrangian method traces the movement of the system in time and space. The Eulerian description (fixed coordinates) uses the concept of the 'control volume', a fixed volume in space, and it monitors the fluid behaviour within the control volume by balancing the fluxes through the surface boundaries with the changes occurring within the control volume.

The dynamics of a continuous medium can be effectively approximated by a sufficiently large number of particles that would at least statistically simulate the behaviour of the continuum.

The description of the properties of the continuum (density, velocity, pressure) is done through the tracking of the position and the properties of the numerous particles that simulate it. Applications of that methodology in fluid mechanics, is known as smoothed particle hydrodynamics (SPH) (Scarlatos and Mehta 1993; Kourafalou et al. 2004; Zafirakou-Koulouris et al. 2012).

The correspondence between the real mass and the mass of the particles is linear. For instance, if 1000 kg of water mass is simulated by 100,000 particles, then each particle represents 0.01 kg of water mass. It is obvious that the bigger the number of particles, the more exact is the description, but also the computational requirements increase substantially.

Simulation of a continuum by particles tracked in Lagrangian coordinates eliminates the problem of the numerical diffusion introduced during the solution of the transport equation. It also offers the ability to track the shape of the pollutant plume at the sub-grid scale (in space smaller than the cell ($\Delta x \Delta y$) used for the description of the hydrodynamics). This methodology applies to a wide range of simulation techniques of discrete particle models and the Monte Carlo probabilistic method.

The movement of dissolved or suspended particles within the fluid involves advective and diffusive processes. The advective motion requires the knowledge of the ambient fluid velocity on the exact location or near the tracked particle. Due to the domain discretization, this information is available either on the nodes or the sides of the grid cells. However, interpolation between the nearest velocity values provides sufficient accuracy.

8.3.1 Application of the random-walk method

The diffusive particle transport process is accomplished by using Einstein's theory for the diffusion phenomena. According to that theory, diffusion is the result of Brownian motion of the mass particulates (molecules) subject to random walks (random velocities) related to the diffusion coefficient. The range of those random velocities of the molecules undergoing diffusion is given as

$$U_B = \sqrt{\frac{6D_t}{t}} \tag{8.18}$$

Thus, the velocity of those particulates is a stochastic variable with a uniform distribution in the range of $-U_B$ to $+U_B$. An illustration of the procedure is schematically given in Figure 8.4.

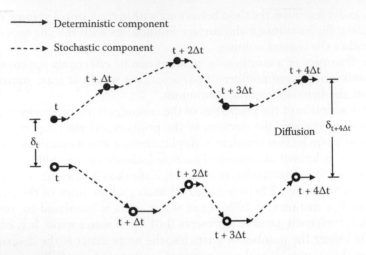

Figure 8.4 Random walk of the particulate matter.

In order to estimate the random velocity, for a certain particle during a certain time step, a random number ranging from –1 to +1 is selected and then multiplied by U_B.

The simulation of a two-dimensional advective and diffusive displacements of K-number of particles (dissolved or suspended) from an initial position $(x_{ok}, y_{ok}, k = 1, ..., K)$ involves the following steps:

1. A deterministic translation with velocities U_d and V_d estimated by interpolation between available neighbouring values of the U, V velocity components of the fluid.
2. A random translation with velocities U_r and V_r selected inside the range $-U_B$ to $+U_B$, by using the relations $U_r = (1 - 2n_{rx})U_B$ and $V_r = (1 - 2n_{ry})U_B$, where n_{rx} and n_{ry} are two random numbers between 0 and 1, internally produced by the computer.
3. The new position of the k-th particle is estimated as $x_k^{n+1} = x_k^n + \Delta x_{dk} + \Delta x_{rk}$ and $y_k^{n+1} = y_k^n + \Delta y_{dk} + \Delta y_{rk}$, where the displacements Δx_{dk}, Δx_{rk}, Δy_{dk}, Δy_{rk} are estimated as the product of the corresponding deterministic or stochastic velocity to time step Δt.
4. A repetition of the procedure for all the particles, taking into consideration the boundary conditions. On a solid boundary, a particle is reflected to its past position; on the end of the flow domain is halted and not displaced anymore; on the free surface is reflected; and on the bed it either re-suspends in the water column or adheres to the bed (erosion and deposition).

The repetition of the procedure for a number of time steps results to the advective and diffusive transport of a diluted or suspended mass in exactly the same way as it would be quantified by solving the transport equation. The concentration in each cell of the discretization grid can be easily measured from the number of particles (representing defined volume or mass of the substance) contained within the specific cell.

For accuracy, this technique requires a large number of tracked particles, so that the concentration in each cell is approximated by a sufficient number of them. In the case of a non-conservative substance, the decomposition or decay of the mass is simulated by removing particles from the moving cluster of particles. The number of particles removed at each time step Δt is related to the decay coefficient λ and the number of the tracked particles as

$$N_{removed} = N_{total} \, \lambda \, \Delta t \qquad\qquad (8.19)$$

This Lagrangian particle-tracking method can be applied for the simulation of either a sudden accidental contaminant release or for a continuous pollution source discharging in receiving waters. In the first case all particles are placed in a specific 'source' location x_o, y_o at $t = 0$. In the second case, a predefined number of new particles are placed at the source location at each time step.

Example 8.3

This application involves the simulation of a contaminant advection and diffusion using a Lagrangian particle-tracking method. The bathymetry, velocity field and location of the contaminant source are the same to the ones in Example 8.1. The source is simulated by 5000 particles introduced at time $t = 0$. The simulation was conducted for a period of 720 time steps (1 hour) and the results are illustrated in Figure 8.5. The similarities between the results of the particle tracking

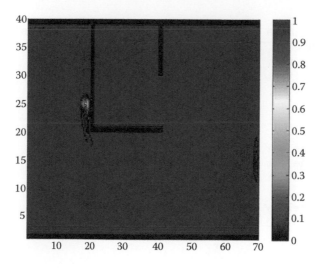

Figure 8.5 Simulation of mass transport using the particle-tracking method.

method (Figure 8.5) and those from the solution of the advection–diffusion equation (Figure 8.2) are evident. The 'entrapment' of particles, shown in the far right side of Figure 8.5, is caused by the boundary conditions. Those particles are related to the long 'tail' of contaminant shown along the horizontal part of the breakwater in Figure 8.2.

Computer code 8.3

```
% Example 8.3 Advective Diffusion by Random Particle Motion
- Marina Breakwater
% dif = Diffusion coefficient;
% dec = Decay coefficient;
% ipp = Number of simulated particles;
% xo, yo = Initial position of particles [m];
% Dd = Spatial computational step [m];
% Dt = Time computational step [s];
% nx = Number of computational steps in the x-direction;
% ny = Number of computational steps in the y-direction;
% nt = Number of computational steps in time;
clc; clear all; close all;
% Input data
dif=0.2;
dec=0.00001;
ipp=5000;
xo=90;
yo=125;
Dd=5;
Dt=5;
nx=70;
ny=40;
nt=720;
% Importing data for depths, velocities and diffusion
coefficients;
load Depths.txt
load CoastU.txt
load CoastV.txt
fDep=Depths;
fCoU=CoastU;
fCoV=CoastV;
% Subroutine for depths h=fm(i,j,fDep);
% Subroutine for velocity u=fm(i,j,fCoU);
% Subroutine for velocity v=fm(i,j,fCoV);
for i=1:nx
    for    j=1:ny
        cmax(i,j)=0;
    end
end
imm=(nx-1)*Dd;
```

```
jmm=(ny-1)*Dd;
% Initial position of particles;
for i=1:ipp
    x(i)=xo;
    y(i)=yo;
end
% Number of decaying particles on every time step;
depp=round(ipp*Dt*dec);
diff=sqrt(6*dif/Dt);
% Main program;
for k=1:nt
    % Movement of particles by advection and diffusion;
    for i=1:ipp
        X=fix(x(i)/Dd)+1;
        Y=fix(y(i)/Dd)+1;
        if X>2 && X<nx-1 && Y>2 && Y<ny-1
            u=fm(X,Y,fCoU);
            ur=fm(X+1,Y,fCoU);
            v=fm(X,Y,fCoV);
            vo=fm(X,Y+1,fCoV);
            uu=(u+ur)/2;
            vv=(v+vo)/2;
            tempx=x(i);
            tempy=y(i);
            ax=rand;
            x(i)=x(i)+(2*(ax-0.5)*diff+uu)*Dt;
            ay=rand;
            y(i)=y(i)+(2*(ay-0.5)*diff+vv)*Dt;
            X=fix(x(i)/Dd)+1;
            Y=fix(y(i)/Dd)+1;
            % Deflection on solid boundarues;
            h=fm(X,Y,fDep);
            if h<0.1
                x(i)=tempx;
                y(i)=tempy;
            end
        end
    end
    % Decaying of particles;
    if dec>0
        for i=1:depp
            ad=rand;
            X=fix(ad*ipp)+1;
            x(X)=0;
            y(Y)=0;
        end
    end
    % Computation of particle concentration in grid cells;
    for i=1:nx
```

```
      for j=1:ny
          c(i,j)=0;
      end
  end
  for i=1:ipp
      X=fix(x(i)/Dd)+1;
      Y=fix(y(i)/Dd)+1;
      c(X,Y)=c(X,Y)+1;
  end
  for j=1:ny
      for i=1:nx
          if c(i,j)>cmax(i,j)
              cmax(i,j)=c(i,j);
          end
      end
  end
  index=k
end
[i,j]=meshgrid(1:1:nx,1:1:ny);
df1=fDep(:,3);
df2=reshape(df1,70,40);
df2=df2';
An=ones(size(df2));
idx=find(df2== 0);
An(idx)=0;
figure;
pcolor(An);
colormap(jet);
shading flat;
hold on;
i=1:nx;
j=1:ny;
Cplot=cmax(i,j)/ipp;
contour(i,j,Cplot')
% colorbar('vert')
```

PROBLEM 8.3

Solve the same problem by making the suggested modifications while keeping the rest of the data constant:

1. Place the contaminant point source at location x = 225 m, y = 190 m.
2. Remove the horizontal arm of the breakwater.
3. Remove the left vertical section of the breakwater.
4. Change the decay coefficient from 0.00001 s^{-1} to (a) 0.01 s^{-1} (decay) and (b) −0.01 s^{-1} (growth).
5. Change the number of simulated particles from 5000 to (a) 500 and (b) 50,000.

Run the simulations, compare the results and comment on the changes observed regarding the spreading of the contaminant plume.

Example 8.4

This application involves the simulation of a contaminant advection and diffusion using a Lagrangian particle-tracking method for the Thermaikos Gulf. The bathymetry, velocity field and location of the contaminant source are the same to the ones in Example 8.2. The source is simulated by 5000 particles introduced at time t = 0. The simulation was conducted for a period of 288 time steps (2 days) and the results are illustrated in Figure 8.6. By comparing the results of the particle-tracking method (Figure 8.6) and those from the solution of the advection–diffusion equation (Figure 8.3), it can be seen that the Eulerian method shows a balance between advective and diffusive transport while the Lagrangian method emphasizes the diffusive effects. However, overall there was agreement between the two methods.

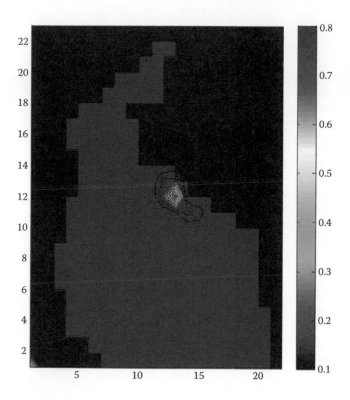

Figure 8.6 Mass transport in Thermaikos Gulf using the particle-tracking method.

Computer code 8.4

```
% Example 8.4 Advective Diffusion by Random Particle Motion
- Thermaikos Gulf
% dif = Diffusion coefficient;
% dec = Decay coefficient;
% ipp = Number of simulated particles;
% xo, yo = Initial position of particles [m];
% Dd = Spatial computational step [m];
% Dt = Time computational step [s];
% nx = Number of computational steps in the x-direction;
% ny = Number of computational steps in the y-direction;
% nt = Number of computational steps in time;
clc; clear all; close all;
% Input data
dif=5;
dec=0.00001;
ipp=5000;
xo=24000;
yo=24000;
Dd=2000;
Dt=600;
nx=22;
ny=23;
nt=288;
% Importing data for depths, velocities and diffusion
coefficients;
load ThermD.txt
load ThermU.txt
load ThermV.txt
fDep=ThermD;
fCoU=ThermU;
fCoV=ThermV;
% Subroutine for depths h=fm(i,j,fDep);
% Subroutine for velocity u=fm(i,j,fCoU);
% Subroutine for velocity v=fm(i,j,fCoV);
for i=1:nx
    for    j=1:ny
        cmax(i,j)=0;
    end
end
imm=(nx-1)*Dd;
jmm=(ny-1)*Dd;
% Initial position of particles;
for i=1:ipp
    x(i,1)=xo;
    y(i,1)=yo;
end
% Number of decaying particles on every time step;
depp=round(ipp*Dt*dec);
```

```
diff=sqrt(6*dif/Dt);
% Main program;
for k=1:nt
    % Movement of particles by advection and diffusion;
    for i=1:ipp
        X=fix(x(i,1)/Dd)+1;
        Y=fix(y(i,1)/Dd)+1;
        if X>2 && X<nx-1 && Y>2 && Y<ny-1
            u=fm(X,Y,fCoU);
            ur=fm(X+1,Y,fCoU);
            v=fm(X,Y,fCoV);
            vo=fm(X,Y+1,fCoV);
            uu=(u+ur)/2;
            vv=(v+vo)/2;
            tempx=x(i,1);
            tempy=y(i,1);
            ax=rand;
            x(i,1)=x(i,1)+(2*(ax-0.5)*diff+uu)*Dt;
            ay=rand;
            y(i,1)=y(i,1)+(2*(ay-0.5)*diff+vv)*Dt;
            X=fix(x(i,1)/Dd)+1;
            Y=fix(y(i,1)/Dd)+1;
            % Deflection on solid boundarues;
            h=fm(X,Y,fDep);
            if h<0.1
                x(i,1)=tempx;
                y(i,1)=tempy;
            end
        end
    end
    % Decaying of particles;
    if dec>0
        for i=1:depp
            ad=rand;
            X=fix(ad*ipp)+1;
            x(X,1)=10;
            y(X,1)=50;
        end
    end
    % Computation of particle concentration in grid cells;
    for i=1:nx
        for j=1:ny
            c(i,j)=0;
        end
    end
    for i=1:ipp
        X=fix(x(i,1)/Dd)+1;
        Y=fix(y(i,1)/Dd)+1;
        c(X,Y)=c(X,Y)+1;
    end
```

```
    for j=1:ny
        for i=1:nx
            if c(i,j)>cmax(i,j)
                cmax(i,j)=c(i,j);
            end
        end
    end
    index=k
end
[i,j]=meshgrid(1:1:nx,1:1:ny);
df1=fDep(:,3);
df2=reshape(df1,22,23);
df2=df2';
An=ones(size(df2));
idx=find(df2== 0);
An(idx)=0;
figure;
pcolor(An);
colormap(jet);
shading flat;
hold on;
i=1:nx;
j=1:ny;
Cplot=cmax(i,j)/ipp;
contour(i,j,Cplot')
% colorbar('vert')
```

PROBLEM 8.4

Solve the same problem by making the suggested modifications while keeping the rest of the data constant:

1. Place the contaminant point source at location x = 16 km, y = 34 km.
2. The source ceases to pollute after 1440 minutes.
3. Replace the decay coefficient with a growth coefficient of the same rate.
4. Place two additional pollutant sources at locations (28 km, 20 km) and (32 km, 16 km).
5. Change the time step to Δt = 120 min, 180 min and 240 min.

Run the simulations, compare the results and comment on the changes observed regarding the spreading of the contaminant plume.

8.4 MECHANICS OF SEDIMENT TRANSPORT

8.4.1 General concepts of sediment transport

The sediment transport phenomena constitute an integral part of inland and coastal hydraulics, since sediment dynamics and sediment quality can cause serious water management, environmental and engineering problems. Aquatic sediments are subject to a dynamic cycle of erosion, transport, settling, deposition and re-suspension. For granular (non-cohesive) sediments, the driving forces are all mechanical, that is, inertial, gravity and impact from particle collisions. For cohesive sediments, in addition to the mechanical forces, electrochemical forces play a predominant role through the processes of aggregation and flocculation.

Sediments are the result of weathering of surface rocks and other surficial geologic formations. Stream sediments can enter the system from upstream sources, known as 'wash load', or can be entrained from bed erosion caused by the shear stresses of the moving water.

The critical shear stress to mobilize and keep granular sediments in suspension is defined by the well-known Shields diagram. This diagram relates the non-dimensional bed shear stress τ^* to the boundary Reynolds number R_e^*, where

$$\tau^* = \frac{\tau_b}{(\rho_s - \rho)gD} \tag{8.20}$$

$$R_e^* = \frac{u_* D}{\nu} \tag{8.21}$$

In Equations 8.20 and 8.21 ρ_s is the sediments density, ρ is the water density, ν the kinematic viscosity of the water and D is a characteristic particle diameter. Furthermore, u^* is the shear velocity (see Chapter 5, Equation 5.37) and τ_b is the bed shear stress, defined, respectively, as

$$u^* = \sqrt{\frac{\tau_b}{\rho}} = \sqrt{gR_h S_e} \tag{8.22}$$

$$\tau_b = \rho f_b u^2 \tag{8.23}$$

According to the Shields data for $1 \le R_e^* \le 10^3$, the critical stress for erosion can be approximated as

$$\tau_{cr}^* = 0.05 \tag{8.24}$$

This relation is also applicable to the case of combined shear effects induced by waves (τ_{bw}) and currents (τ_{bc}). In that case the bed shear is assumed to be the sum of the two stresses ($\tau_b = \tau_{bc} + \tau_{bw}$). This is valid for co-linear currents and waves. In all other cases the sum is vectors sum.

Sediments are transported near the bed as 'bed load' (q_b) (by saltation, rolling or inter-collision) or within the water column as 'suspended load' (q_s). The separation between these two transportation modes is artificial, since in reality there is a continuous sediment particle exchange between bed load and suspended load.

The rate of transported sediments is expressed in terms of the volume, mass or weight of sediment material carried per unit width (specific discharge). The 'total load' (q_t) can be obtained by adding the bed load and the suspended load (Yang 2003).

The concentration of suspended sediments is not uniform along the depth $C(z)$ (concentration increases from the water surface towards the bed). Thus, in a horizontal two-dimensional domain, under steady-state conditions, the suspended load can be estimated as

$$q_{sx} = \int_{\alpha}^{h} uC(z)\,dz \quad \text{and} \quad q_{sy} = \int_{\alpha}^{h} vC(z)\,dz \qquad (8.25)$$

where h is the water depth and α is the thickness of the bed load layer.

For more than a century now, extensive theoretical and experimental research is being conducted in order to understand the processes involved and to quantify the specific discharges of bed load, suspended load or total load in terms of flow characteristics, such as flow velocity, water depth, eddy diffusion coefficient in the vertical direction, and the wave period and height in the case of coastal transport (Fredshoe and Deigaard 1992).

8.4.2 Quantification of sediment transport

In a horizontal two-dimensional domain, the rate of change of the water depth, h, or of the altitude of the bed elevation, ζ_b, both measured from some reference datum, can be calculated by considering the specific sediment discharges:

$$\frac{\partial h}{\partial t} = -\frac{\partial \zeta_b}{\partial t} = \frac{\partial q_{tx}}{\partial x} + \frac{\partial q_{ty}}{\partial y} \pm S_s \qquad (8.26)$$

where S_s is a source (dumping) or sink (excavation) term. Note that the specific discharges are in volumetric units and include the voids.

A large number of empirical sediment transport formulas is available. However, since those formulas were derived using different laboratory and field data, the results that they produce may vary significantly. In addition, a common characteristic of all those formulas is their non-linearity, that is, the dependence of the specific sediment discharges to a high power (3 to 5) of the hydrodynamic variables (e.g. flow velocity). This fact causes large estimate variations of the sediment discharge even for a small change of the water velocities, leading to further uncertainty in the prediction of sediment discharges and consequent bed deformation.

8.4.2.1 Meyer-Peter and Muller formula
for bed load transport

An example of a nonlinear relation is the classical Meyer-Peter and Muller formula for bed sediment discharge that was derived using energy gradient concepts. The equation written in metric units reads

$$q_b^{2/3} \left[\frac{(\rho_s - \rho)}{\rho_s} \right]^{2/3} \frac{\rho^{1/3}}{4g(\rho_s - \rho)D} = \left(\frac{k_s}{k_r} \right)^{3/2} \frac{\rho R_h S_e}{(\rho_s - \rho)D} - 0.47 \qquad (8.27)$$

where q_b is the bed load, ρ_s and ρ are the sediment and water densities, D is the mean particle diameter, R_h is the hydraulic radius and S_e is the energy gradient. The ratio $\dfrac{k_s}{k_r}$ defines the ratio of the energy required for sediment motion over the total energy; k_s is the combined bed friction and k_r is the skin friction estimated as

$$k_r = \frac{26}{D_{90}^{1/6}} [m^{1/3}s] \qquad (8.28)$$

The particle diameter D_{90} denotes that 90% of the material is finer. In the case of no bed forms, the ratio varies $0.5 \le \dfrac{k_s}{k_r} \le 1.0$.

8.4.2.2 Engelund and Hansen method
for total load sediment transport

One of the approaches developed for estimation of the total sediment transport, q_t, based on the Bagnold's stream power concept, is the Engelund and Hansen formula

$$\frac{q_t}{\sqrt{\gamma_s \left(\frac{\gamma_s - \gamma}{\gamma} \right) gD^3}} = 0.05 \frac{u^2}{ghS_e} \left[\frac{\tau}{(\gamma_s - \gamma)D} \right]^{5/2} \qquad (8.29)$$

The total discharge is estimated in terms of sediment weight transported. From Equation 8.29 the non-linearity of the total sediment load in terms of hydraulic characteristics is evident.

Overall, the sediment transport problem becomes very complex if the desired operational goal is to predict the long-term evolution of the bed morphology due to continuous sediment transport processes. Even in the simplest case of a one-dimensional moveable bed canal with sediments transported under the action of the flowing water, a highly non-linear hyperbolic equation may be required to describe the phenomenon. Under steady-state uniform flow conditions, the water discharge across the canal per unit canal length, q_w, is constant:

$$q_w = uh = constant \qquad (8.30)$$

Therefore, in order for Equation 8.30 to remain constant, the mean velocity (u) and the average depth (h) would change accordingly. Based on the Meyer-Peter and Muller formula, the bed load sediment discharge can be expressed to the flow velocity (u) as

$$q_b = \alpha u^3 + \beta \qquad (8.31)$$

where α and β are coefficients depending on the sediment characteristics (Samaras and Koutitas 2008). Utilizing mass continuity principles for the movable bed, the bed evolution can be described by any of the following equations:

$$\frac{\partial \zeta_b}{\partial t} = -\frac{\partial q_b}{\partial x} \qquad (8.32)$$

$$\frac{\partial h}{\partial t} = \alpha \frac{\partial u^3}{\partial x} \qquad (8.33)$$

$$\frac{\partial h}{\partial t} = \alpha q_w^3 \frac{\partial h^{-3}}{\partial x} = -\frac{3\alpha q_w^3}{h^4}\frac{\partial h}{\partial x} = -c^* \frac{\partial h}{\partial x} \qquad (8.34)$$

Equation 8.34 is a advective transport equation (see Chapter 3, Equation 3.30), where the 'signal' h(x,t) propagates with a celerity c* along the x-direction.

Example 8.5

This application simulates the evolution of a dredged section in a moveable bed canal by utilizing the bed load sediment transport equations (Equations 8.31 and 8.32). The data used are as follows:

> Length of canal section = 400 m
> Water depth = 2 m
> Water discharge per unit width = 2 m³/m/s
> Sediment transport coefficient = 0.01 s²/m
> Location of dredged section = Centre of dredge is 102 m from the canal entrance
> Dredge geometry = Trapezoidal, bottom width 16 m, top width 52 m, depth 2 m below the bed

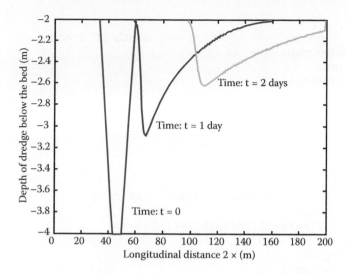

Figure 8.7 Propagation and deformation of a dredged section within a canal.

The domain was discretized into 200 segments ($\Delta x = 2$ m) with a time step $\Delta t = 20$ s. The simulation lasted for 2 days, and the results of the propagation and deformation of the dredged section after 1 day and at the end of the simulation are shown in Figure 8.7.

Here it should be noted that the numerical solution is very sensitive to the sediment transport coefficient, a. From the results it is evident that the upwind finite differences solution scheme introduces numerical diffusion that flattens the steepness of the propagating bed and also that it causes a slight deviation from the volume (area) conservation.

Computer code 8.5

```
% Example 8.5 Nonlinear Sediment Transport Along a Dredged
Canal
% ho = Initial water depth [m];
% qw = Water discharge per unit width [m^3/m/s]
% a = Sediment transport coefficient [s^2/m];
% Dx = Spatial step [m];
% Dt = Time step [s];
% nx = Number of spatial steps;
% nt = Number of temporal steps;
% Input data;
a=0.01;
ho=2;
qw=2.0;
Dx=2;
Dt=20;
```

```
nx=200;
nt=8640;
% Canal topography - Location and depth of dredge;
load Dredge.txt
for i=1:nx
    h(i)=Dredge(i,2);
end
% Initialization of variables;
for i=1:nx
    hn(i)=h(i);
end
% Main program;
for k=1:nt
    % Sediment transport and bed reshaping;
    for i=2:nx
        q(i)=a*qw^3/(h(i)+h(i-1))^3;
    end
    q(1)=q(2);
    for i=2:nx-2
        if q(i)+q(i+1)>0
            adv=(q(i)-q(i-1))/Dx;
        else
            adv=(q(i+2)-q(i+1))/Dx;
        end
        hn(i)=h(i)+Dt*adv;
    end
    hn(1)=hn(2);
    hn(nx-1)=hn(nx-2);
    for i=2:nx-2
        h(i)=hn(i);
    end
    h(1)=h(2);
    h(nx-1)=h(nx-2);
    % Storage of depths at time=0;
    if k==1
        for i=1:nx
            ho(i)=h(i);
        end
    end
    % Storage of depths at half of the total simulation
    time;
    if k==nt/2
        for i=1:nx
            h1(i)=h(i);
        end
    end
end
i=1:nx;
hoplot=-ho(i);
h1plot=-h1(i);
```

```
hplot=-h(i);
plot(i,hoplot,'','Linewidth',1.5)
hold on
plot(i,h1plot,'r','Linewidth',1.5)
plot(i,hplot,'g','Linewidth',1.5)
xlabel('Longitudinal distance 2 x [m]')
ylabel('Depth of dredge below the bed [m]')
text(58,-3.8,'Time: t = 0')
text(75,-3.0,'Time: t = 1 day')
text(120,-2.6,'Time: t = 2 days')
```

PROBLEM 8.5

Solve the same problem by making the suggested modifications while keeping the rest of the data constant:

1. Change the water discharge from 2 m³/m/s to 0.5, 1.0, 1.5 and 2.5 m³/m/s.
2. Change the sediment transport coefficient from 0.01 s²/m to 0.001, 0.005 and 0.15 s²/m.
3. Change the distance from the centre of the dredge to the canal entrance to 62 m, and change the shape to trapezoidal with bottom width 4 m, top width 16 m and a depth 1 m below the bed.
4. Put two dredges of the size and shape of part (c) with their centres at 120 and 280 meters.
5. Change the exponent of the bed load (Equation 8.31) from 3 to 2.5 and 3.5.

Run the simulations, compare the results and comment on the changes observed regarding the movement and shape of the dredge.

Example 8.6

This application qualifies the scouring/deposition processes around a breakwater structure in a horizontal two-dimensional domain. The bathymetry and the velocities field are the same to those used in Example 8.1. Other data used for the model are as follows:

> Mean particle diameter = 0.001 m
> Sediment density = 2500 kg/m³
> Bed friction coefficient = 0.001
> Spatial step = 5 m
> Time step = 1 s

The model is based on the sediment transport continuity equation (Equation 8.26) and the Engelund-Hansen formula (Equation 8.29). The simulation started with a flat bed and deformed under the

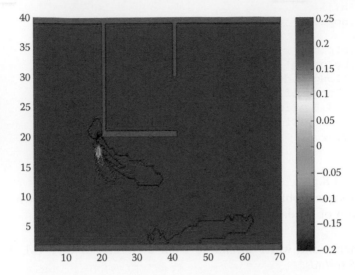

Figure 8.8 Sediment bed scouring and accretion due to water currents.

influence of current velocities. After a simulation period of 120 days it can be seen (Figure 8.8) that at the corner of the breakwater, where current velocities are high, there is local scouring, while adjacent to the scouring the eroded sediments were re-deposited. A close-up of the sedimentation phenomenon is presented in Figure 8.9.

It should be emphasized that although the model depicts the location of the bed deformation, the predicted scouring and accretion depths are very approximate and require calibration of the physical

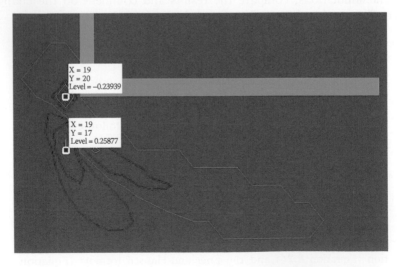

Figure 8.9 Close-up illustration of the sediment scouring and deposition effects.

parameters from actual field data. The ability of a good qualitative description but weak quantitative one is characteristic of sediment transport models.

Computer code 8.6

```
% Example 8.6 Sediment Transport with Engelund-Hansen Method
% d = Particle diameter [m];
% fb = Bed friction coefficient;
% dr = Relative density difference;
% dwtr = Water density [kg/L];
% dsed = Sediment density [kg/L];
% Dd = Spatial step (same in both directions)[m];
% Dt = Time step [s];
% nx = Number of spatial steps in the x-direction;
% ny = Number of spatial steps in the y-direction;
% nt = Time of simulation steps [s];
clc; clear all; close all;
% Input data;
g=9.81;
d=0.001;
fb=0.001;
dwtr=1.0;
dsed=2.65;
dr=(dsed-dwtr)/dwtr;
Dd=5;
Dt=1;
nx=70;
ny=40;
nt=10368000;
up2=dr*g*d;
qss=d*sqrt(up2);
% Importing data for depths, and velocities;
load Depths.txt
load CoastU.txt
load CoastV.txt
fDep=Depths;
fCoU=CoastU;
fCoV=CoastV;
% Subroutine for depths h=fm(i,j,fDep);
% Subroutine for velocity u=fm(i,j,fCoU);
% Subroutine for velocity v=fm(i,j,fCoV);
% Initialization of variables;
for i=1:nx
    for j=1:ny
        qsx(i,j)=0;
        qsy(i,j)=0;
        dh(i,j)=0;
    end
```

```
end
% Main program;
for j=2:ny-2
    for i=2:nx-2
        h=fm(i,j,fDep);
        if h>0.1
            u=fm(i,j,fCoU);
            tbx=1000*fb*u^2;
            tbxs=tbx/(1000*up2);
            qsx(i,j)=0.05*tbxs^2.5/fb*qss;
            v=fm(i,j,fCoV);
            tby=1000*fb*v^2;
            tbys=tby/(1000*up2);
            qsy(i,j)=0.05*tbys^2.5/fb*qss;
        end
    end
    for j=1:ny
        qsx(nx-1,j)=qsx(nx-2,j);
        qsy(nx-1,j)=qsy(nx-2,j);
    end
end
for j=2:ny-2
    for i=2:nx-2
        dh(i,j)=(qsx(i+1,j)-qsx(i,j))/Dd+(qsy(i,j+1)-
        qsy(i,j))*Dt/Dd;
    end
end
for j=1:ny
    for i=1:nx
        DD(i,j)=dh(i,j)*nt;
        % Avoid plotting trivial sediment erosion/deposition
        effects;
        if abs(DD(i,j))<0.01;
            DD(i,j)=0;
        end
    end
end
[i,j]=meshgrid(1:1:nx,1:1:ny);
df1=fDep(:,3);
df2=reshape(df1,70,40);
df2=df2';
An=ones(size(df2));
idx=find(df2== 0);
An(idx)=0;
figure;
pcolor(An);
colormap(jet);
shading flat;
hold on;
i=1:nx;
```

```
j=1:ny;
dhplot=DD(i,j);
% figure
contour(i,j,dhplot')
% colorbar('vert')
```

PROBLEM 8.6
Solve the same problem by making the suggested modifications while keeping the rest of the data constant:

1. Change the sediment particle diameter from 0.001 m to 0.002, 0.003 and 0.0005 m.
2. Change the sediment density from 2.65 kg/m³ to 1.65 and 3.65 kg/m³.
3. Change the bed friction coefficient from 0.001 to 0.005, 0.0005 and 0.00025.
4. Increase the magnitude of the velocity field by 20%.
5. Reduce the water depth of the entire domain by 25%.

Run the simulations, compare the results and comment on the changes observed regarding the extent and shape of the bed scouring and shoaling.

8.4.3 Alongshore sediment transport

A very important management tool for coastal zones is the one-line alongshore sediment transport model (Pelnard-Considère 1956). This model is based on the volumetric conservation principle coupled with an empirical sediment transport formula. Experimentally it has been found that the alongshore transport formula can be written in the form

$$Q_s = a_s H^m \sin(2\varphi)$$ (8.35)

where Q_s is the volumetric rate of transported sediments along the coast and inside the breaker zone, H is the wave height, φ is the wave breaking angle with respect to the coastline, m is an exponent varying between 2.0 and 3.5 and a_s is an empirical coefficient incorporating the sediment particle diameter, beach slope and wave period (Kamphuis 1991).

8.4.3.1 One-line model

By lumping all of the three dimensional effects in the distribution of $Q_s(x,t)$, the evolution of the coastline can be mathematically described by the one-line equation as

$$\frac{\partial y}{\partial t} = \frac{1}{h_{max}} \frac{\partial Q_s}{\partial x}$$ (8.36)

where $y(x,t)$ is the shoreline retreat (erosion) or advancement (accretion) perpendicularly to the original shoreline, and h_{max} is the maximum water depth that sediment transport is affected. That depth is usually related to the wave breaking height ($h_{max} = 2.5H_b$).

Equations 8.35 and 8.36 are solved for the two unknown variables $y(x,t)$ and $Q_s(x,t)$ while the equations feed back to each other as the wave breaking angles change in time from the initial distribution $\varphi_o = \varphi(x, t = 0)$ as

$$\varphi(x,t) = \varphi_o - \arctan\left(\frac{dy}{dx}\right) \tag{8.37}$$

The angle φ is measured counterclockwise. In practice the alongshore sediment transport process can be modified/interrupted due to various reasons such as

- Presence of a groin interrupting the sediment flow
- Presence of a detached breakwater creating a wave 'shadow' on the coast and thus changing the values of the Q_s, h_{max} and φ
- Unavailability of sediment source, etc.

The numerical solution of the system of Equations 8.35 to 8.37 is done by using a centred finite differences scheme on a staggered discretization one-dimensional grid. Thus, the coastline is divided into equal-length segments and the retreat or extension of the beach along the y-axis (seaward) is calculated at the centre of the segment from the sediment fluxes defined at the sides of the segment.

Example 8.7

This application investigates the evolution of an initially straight coastline around a groin by using the one-line model coupled with the alongshore sediment transport formula. The beach is subject to a continuous attack from incoming waves approaching the beach at an angle φ. The data used for the simulation are as follows:

Reference water depth = 2.5 m
Sediment transport coefficient (Equation 8.35) = 0.02
Empirical exponent (Equation 8.35) = 2.5
Wave heights = 1.0 m (modified near the groin as in the input file)
Angle of incident wave = 0.5 radians (28.66°) (modified near the groin as in the input file)
Initial beach profile = Straight line
Location of the groin = 500 m (centre of the solution domain)
Length of the beach = 1000 m

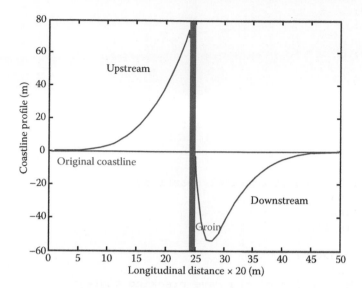

Figure 8.10 Deformation of a beach profile around a groin.

The beach was discretized into 50 segments of $\Delta x = 20$ m each and the time step used was $\Delta t = 360$ s. The deformation of the beach after a period of 30 days is shown in Figure 8.10. The resulted profile is very typical with those found in beaches where sediment deposits (accretion) in the direction of incoming waves and erodes (scouring) in the leeway of the groin (or jetty).

Computer code 8.7

```
% Example 8.7 Coastal Evolution Subject to Longshore
Sediment Transport
% H = Reference water depth [m];
% at = Sediment transport coefficient;
% m = Empirical exponent [2 < m < 3.5];
% kk = Wave height magnitude modification coefficient;
% mm = Wave breaking angle magnitude modification
coefficient;
% Dx = Spatial step [m];
% Dt = Time step [s];
% nx = Number of spatial steps;
% nt = Number of time steps;
clc; clear all; close all;
% Input data;
H=2.5;
at=0.02;
m=2.5;
```

```
kk=1.0;
mm=1.0;
Dx=20;
Dt=360;
nx=50;
nt=7200;
% Wave height data file;
load a.mat
% Angle of incident wave data file;
load fo.mat
% Location of the groin;
load N.mat
% Configuration of original coastline;
load yo.mat
for i=1:nx
    y(i)=yo(i);
end
% Main program;
for k=1:nt
    % Computation of the wave breaking angle;
    for i=2:nx-1
        f(i)=mm*fo(i)-atan((y(i)-y(i-1))/Dx);
    end
    % Computation of the longshore transport;
    for i=2:nx-1
        if N(i) ~=1
            q(i)=at*(kk*a(i))^m*sin(2*f(i));
        end
    end
    q(1)=2*q(2)-q(3);
    q(nx)=2*q(nx-1)-q(nx-2);
    % Estimation of the new coastline;
    for i=1:nx-1
        y(i)=y(i)-Dt/Dx*(q(i+1)-q(i))/(2*H);
        % Consideration of the coast (Groin; Breakwater;
        Coastline)
        if N(i)== 2 && y(i)<yo(i)
            y(i)=yo(i);
        elseif N(i)== 3 && y(i)>yf
            y(i)=yf;
        end
    end
end
par=find(N== 1);
x_groin=par-0.4;
i=1:nx;
yplot=y(i);
plot(i,yplot,'b','Linewidth',1.5)
text(35,-30,'downstream')
text(10,50,'upstream')
```

```
xlabel('Longitudinal distance x 20 [m]')
ylabel('Coastline profile [m]')
% subroutibe "hline";
h=hline(0,'b','Original coastline');
% subroutine "vline";
h=vline(x_groin,'m','Groin');
```

PROBLEM 8.7

Solve the same problem by making the suggested modifications while keeping the rest of the data constant:

1. Change the sediment transport coefficient from 0.02 to 0.1, 0.01, 0.001 and 0.0001.
2. Change the empirical exponent from 2.5 to 2.0, 3.0 and 3.5.
3. Change the wave height by a factor of 0.5, 1.5 and 2.0.
4. Change the wave breaking angle by a factor of 0.5, 1.2, 1.5 and 1.7.
5. Modify the code so that at the transport coefficient changes linearly from 0.02 at the beginning of the simulation to 0.0002 at the end of the simulation.

Conduct the simulations, compare and explain the results, and comment on the observations regarding the development of the upstream aggradation and the downstream erosion of the coastline.

8.4.4 Cross-shore sediment transport by waves

8.4.4.1 Conceptual description

The development of alongshore sandbars is a very important physical process resulting from onshore–offshore sediment movement by waves in the wave breaking zone. Alongshore sandbars can act as a natural coastal defence against winter waves, that is, those with high steepness (H/L), and consequently with the strongest erosive capacity. Qualitatively the process can be described as follows:

1. In the zone offshore of the wave breaking line, the waves transport sediment as bed load towards the coast, while in the mid-depth water layer, sediments are transported offshore as suspended sediment.
2. Inside the breaker zone the waves take the form of a sequence of quasi-solitary waves, transporting water masses towards the coast, and resulting to a mean-water-surface setup, $\Delta\eta$, related to the wave height and period.
3. The transported water masses return to the sea via a strong near-bed current, the 'undertow', which also transports sediments as bed load towards the open sea.

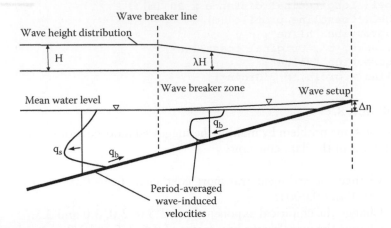

Figure 8.11 Sediment transport in the wave breaker zone.

4. The two near-bed sediment fluxes, one moving onshore outside the breaking line and one moving offshore inside the breaking line, converge on the breaking line, where the sediment is accumulated, creating in time a sand bar.
5. This bar increases to a shape that is stabilized due to the sediment transport threshold and to the maximum possible natural bed slope (angle of repose).
6. This bar locally reduces the water depth and consequently the breaking wave height from this position to the coastline, and thus reduces the erosive capacity of the waves.

In summary, sand masses are transported through erosion, mainly from the inshore region towards the wave breaking line, thus forming a defence from further beach erosion.

The mathematical modelling of the process aims to the quantitative description of the interacting processes of wave propagation in the breakers zone, the simultaneous transport of sediment to and from the coast, and the description of the evolution of the bed geometry, from an initial straight line to a curved one, with eroded and accreted segments. The process and the basic notations are illustrated in Figure 8.11.

8.4.4.2 Mathematical formulation

Detailed description of the phenomenon including wave propagation, period-averaged transport of water masses and simultaneous transport of sediment coupled (feed-backing) with the hydrodynamic characteristics in the area (wave height distribution and wave induced currents) is extremely complicated. In the following, a mathematical model will be developed by using some well-established quantitative relations in combination with the principle of mass conservation for the sediment in the cross-shore direction. The result will be a partial differential equation (PDE) in terms of the water depth h(x,t) across the beach.

The wave height, H, is assumed constant in the region outside the breaker zone. The wave induced current, U_c, (period-averaged) near the bed is mostly due to the Stokes drift, that is, due to the open trajectories of the water particles:

$$U_c = \left(\frac{\pi H}{L}\right)^2 \frac{c_o}{2\sinh^2(kh)} \tag{8.38}$$

where L is the wave length, c_o is the wave celerity, h is water depth and $k = \frac{2\pi}{L}$ is the wave number. In addition, the amplitude of the near-the-bed orbital wave velocity, U_o, can be defined by the first-order wave theory as

$$U_o = \frac{\pi H}{T} \frac{1}{\sinh(kh)} \tag{8.39}$$

The water depth at the breaking line is related to the wave height, H,

$$h_b = \frac{H}{0.8} \tag{8.40}$$

and the breaking wave height is related to the water depth at the breaking line

$$H_b = h_b\lambda \tag{8.41}$$

where

$$\lambda = \xi^{0.17} + 0.08 \tag{8.42}$$

$$\xi = \frac{\tan\theta}{\sqrt{\dfrac{H}{L}}} \tag{8.43}$$

with θ the beach slope.

The water depth inshore of the break line is linearly increased (in relation to the still water depth) up to the coastline, where the wave setup Δη is added, estimated as

$$\Delta\eta = \frac{H}{1 + \dfrac{8}{3\lambda^2}} \tag{8.44}$$

Inside the breaker zone the wave height is linearly changed from the H_b value to the value $\lambda \Delta \eta$ on the coastline. In addition the offshore water flux due to the undertow can be approximated either as (O'Connor et al. 1998)

$$Q = \frac{\pi H^2}{4T} \frac{1}{\tanh(kh)} + 0.9 \frac{H^2}{T} \tag{8.45}$$

or

$$Q = \kappa \frac{H^2}{T} \tag{8.46}$$

where $O(\kappa) = 2 \div 3$. This flux (or specific discharge) is transported offshore in a water column of height equal to 0.8h (the space under the trough of the breaking waves). The resulting current velocity, U_c, is

$$U_c = \frac{Q}{0.8h} \tag{8.47}$$

The amplitude of the orbital water velocity, U_o, can be described using the long-waves theory as

$$U_o = \frac{H}{2h} \sqrt{gh} \tag{8.48}$$

From the pairs of the velocity values U_c and U_o, inside and outside the break line, the combined bed shear due to wave motion and the wave induced current is

$$\frac{\tau_{wc}}{\rho} = \frac{gU_c^2}{C_h^2} + 0.25 f_w U_o^2 \tag{8.49}$$

where C_h and f_w are the bed friction coefficients due to currents and waves, respectively, defined by

$$C_h = 18 \log\left(\frac{12h}{\varepsilon}\right) \tag{8.50}$$

where ε is the absolute bed roughness, and

$$\ln f_w = -5.99 + 5.12 \left(\frac{\xi_w}{\varepsilon}\right)^{-0.914} \tag{8.51}$$

with ξ_w being the amplitude of the horizontal wave orbital motion. A mean value of the friction coefficient f_w is approximately equal to 0.05.

Based on the Bagnold's 'power approach' concept, various sediment transport models have been developed that combine the action of both waves and currents (Bailard 1981; Camemen and Larsen 2005). For computational simplicity, an approach is applied which relates the total sediment volume transport flux to the magnitudes of mean current and the total (due to waves and currents) bed shear using an extended form of the Engelund-Hansen formula (Engelund and Hansen 1972):

$$Q_s = \frac{0.05 U_o C_h \left(\dfrac{\tau_{wc}}{\rho} \right)^2}{g^{2.5} \left(\dfrac{\rho_s - \rho}{\rho} \right)^2 D_{50}} \tag{8.52}$$

The region between the selected offshore location (close to the break line) and the initial coastline is discretized into intervals of length Δx and the sediment conservation equation is applied numerically in the form

$$\Delta h_i = \frac{\Delta t}{\Delta x} (Q_{s,i} - Q_{s,i-1}) \tag{8.53}$$

leading to the evolution of the water depth with time. The water depth is continuously corrected and the hydrodynamic characteristics are continuously modified in an interactive process.

The weakness of the model is that it does not take into account a sediment incipient motion threshold, permitting the uninterrupted evolution of the bed. Also the model does not account for the bed surface slopes when becoming steeper than the angle of repose, beyond which local sloughing needs to be imposed.

Example 8.8

This application investigates the re-shaping of beach slope and the formation of a sandbar due to incident waves normal to the beach. The model accounts for the wave breaking, wave set, and the period-averaged wave-induced current motion. The resulting sediment transport is quantified by the Engelund-Hansen equation (Equation 8.52). The data used for the simulation are as following:

Water depth at the sea boundary = 5 m
Wave height at the sea boundary = 2 m
Wave period = 6 s
Beach bed slope = 0.02
Bed roughness = 0.001 m
Absolute bed roughness = 0.001 m
Sediment density = 2500 kg/m³

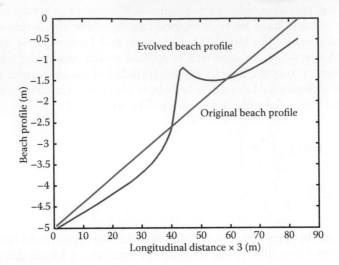

Figure 8.12 Beach profile and the development of a sandbar.

Water density = 1000 kg/m³
Mean particle diameter = 0.001 m

The spatial discretization length is set as $\Delta x = 3$ m and the time step as $\Delta t = 12$ s. After the continuous wave action for a 5-hour period, the deformed beach profile and the evolution of the water depth over the sandbar are shown in Figures 8.12 and 8.13, respectively.

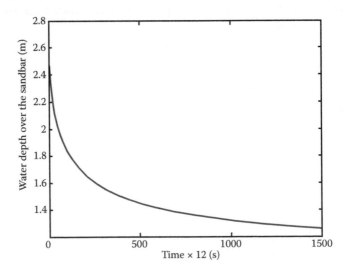

Figure 8.13 Reduction of the water depth due to the development of a sandbar.

Computer code 8.8

```
% Example 8.8 Cross Shore Sediment Transport and Beach
Profile Formation
% H = Maximum water height at the left-hand boundary [m];
% ao = Wave height at the left-hand boundary [m];
% T = Wave period [s];
% S = Bed slope;
% dsed = Sediment density [kg/m^3];
% dwtr = Water density [kg/m^3];
% d50 = Mean particle diameter [m];
% ks = Absolute bed roughness [m];
% h() = Water depth [m];
% Dx = Spatial step [m];
% Dt = Time step [s];
% nt = Number of time steps;
clc; clear all; close all;
% Input data;
g=9.81;
H=5;
ao=2;
T=6;
S=0.02;
ks=0.001;
dsed=2.5;
dwtr=1.0;
d50=0.001;
Dx=3;
Dt=12;
nt=1500;
lo=g*T^2/(2*pi);
sed=g^2.5*((dsed/dwtr)-1)^2*d50;
% Definition of initial depths and solution domain size;
kx=1000;
for k=2:kx
    h(k)=H-S*(k-1)*Dx;
    if h(k)>0
        nxx=k;
    else
        nx=nxx-1;
        break
    end
end
% Breaking water depth;
hb=ao/0.8;
% Estimation of water setup and modified depths;
xsi=S/sqrt(ao/lo);
G=xsi^0.17+0.08;
set=hb/(1+8/3/G^2);
% Estimation of the breaking line location;
```

```
h(1)=H;
for i=2:nx
    h(i)=H-S*(i-1)*Dx;
    if h(i)<hb && h(i-1)>hb
        nb=i;
        break
    end
end
% New depths inside the breaker zone increased by setup;
for i=nb:nx
    h(i)=h(i)+set*(i-nb)/(nx-nb);
end
h(nx+1)=2*h(nx)-h(nx-1);
% Main program;
for k=1:nt
    hb=h(nb);
    % New wave breaking height at nb after the increase of
    water depth hb;
    ab=hb*G;
    % New wave distribution inside breaker zone;
    for i=1:nx
        if i<nb
            a(i)=ao;
        else
            a(i)=ab-(ab-set*G)*(i-nb)/(nx-nb);
        end
    end
    a(nx+1)=2*a(nx)-a(nx-1);
    % Estimation of wave characteristics along domain;
    temp=0;
    for i=1:nx
        l(i)=lo;
        while(abs(l(i)-temp)>0.1)
            temp=l(i);
            arg=2*pi*h(i)/l(i);
            l(i)=lo*(exp(arg)-exp(-arg))/
            (exp(arg)+exp(-arg));
            test=abs(l(i)-temp);
        end
    end
    l(nx+1)=2*l(nx)-l(nx);
    % Orbital and residual velocities; Shear stresses;
    Sediment discharges;
    for i=1:nx
        hmm=(h(i)+h(i+1))/2;
        amm=(a(i)+a(i+1))/2;
        lmm=(l(i)+l(i+1))/2;
        if i>nb
            arg=2*pi*hmm/lmm;
            uc(i)=-2.5*amm^2/T/(0.8*hmm);
```

```
            uo(i)=pi*amm/T/(exp(arg)-exp(-arg))*2;
        else
            arg=2*pi*hmm/lmm;
            cc=lmm/T;
            uc(i)=(pi*amm/T/(exp(arg)-exp(-arg))*2)^2/2/cc;
            uo(i)=pi*amm/T/(exp(arg)-exp(-arg))*2;
        end
        ch=18*log(12*hmm/ks);
        twc=g*uc(i)^2/ch^2+0.25*0.05*uo(i)^2;
        qs(i)=0.05*uc(i)*ch*twc^2/sed;
    end
    % Continuity equation for estimation of new depths;
    for i=2:nx
        hhb(i)=(qs(i)-qs(i-1))/Dx*Dt;
    end
    % Filtering the water depths by moving averages;
    for i=1:nx
        h(i)=h(i)+hhb(i);
    end
    for i=2:nx-1
        hm(i)=0.8*h(i)+0.1*(h(i-1)+h(i+1));
    end
    for i=2:nx-1
        h(i)=hm(i);
    end
    h(nx+1)=2*h(nx)-h(nx-1);
    %    index=k
    hbb(k)=h(nb);
    index=k
end
% Original beach profile;
x=0;
for k=1:83
    x=x+k;
    yplot(k)=-H+0.06*k;
end
ii=1:k;
i=1:nx;
hplot=-h(i);
plot(1:nx,hplot,'Linewidth',1.5)
xlabel('Longitudinal distance x 3 [m]')
ylabel('Beach profile [m]')
hold on;
plot(ii,yplot,'m','Linewidth',1.5)
text(30,-0.7,'Evolved beach profile')
text(50,-2.3,'Original beach profile')
figure
k=1:nt;
hbplot=hbb(k);
plot(1:nt,hbplot,'Linewidth',1.5)
```

```
xlabel('Time x 12 [s]')
ylabel('Water depth over the sandbar [m]')
```

PROBLEM 8.8
Solve the same problem by making the suggested modifications while keeping the rest of the data constant:

1. Change the wave height from 2.0 m to 1.0, 1.5 and 2.5 m.
2. Change the wave period from 6 s to 3, 8 and 9 s.
3. Change the bed slope from 0.02 to 0.018 and 0.022.
4. Change the bed roughness from 0.001 m to 0.0005, 0.0015 and 0.005 m.
5. Change the mean particle diameter from 0.001 m to 0.00075, 0.0015 and 0.005 m.

Run the simulations, analyse the data and comment on the sensitivity of the model to the various parameters involved.

8.5 TRACER METHOD APPLIED TO ACTIVE PARTICLES

8.5.1 Qualitative description

In Section 8.3, it was presented how the numerical diffusion error was eliminated from the simulation of advection–diffusion transport of a substance by using the Lagrangian particle-tracking method. In that case the pre-existing hydrodynamics remained unaffected, a fact characteristic of 'passive' particles, and the movement of the dissolved or suspended matter was described by a large number of such particles.

However, whenever dealing with 'active' particles, the particles drive and interact with the hydrodynamics of the flow domain and the problem becomes a two-phase flow. Typical examples of two-phase flows are hyperconcentrated sediment and debris flows, and flow driven by air bubbles.

In the following, the problem of air-bubble flow will be addressed. Understanding of the air-bubble flow mechanics is of direct interest to a variety of engineering applications such as circulation and aeration in sewage treatment plants, a bubble curtain for salinity control in waterway gate locks and water quality enhancement in small enclosed water bodies.

The two basic parameters for air-bubble flow are the driving air–gas discharge, q_s, and the bubble diameter, D_b (or radius, R_b). It is assumed that the air bubbles soon reach an equilibrium state with the ambient fluid and maintain a vertical velocity relative to the fluid, U_b, which depends on their diameter. The flow is generated by the distributed load of the local drag forces exercised by the ensemble of the bubbles in a unit volume of the fluid. That results in an upward force which interacts with the diffusion process, the pressure gradients and the acceleration of the surrounding fluid mass.

8.5.2 Mathematical formulation

For a two-dimensional vertical domain (the third dimension is considered as unit width) the governing equations are the continuity equation and the two components of the momentum equation, written for the non-dimensional pressure (p) and the two local velocities U and V. The model also incorporates two special features:

- The concept of fluid pseudo-compressibility
- The effect of the almost incompressible air volume introduced or subtracted from the control volume containing water

Based on this, the model equations can be effectively written in terms of surrogate Lagrangian particles directly correlated to the air-bubble flow characteristics. Thus the continuity equation reads as

$$\frac{\partial p}{\partial t} + \frac{\partial U}{\partial x} + \frac{\partial V}{\partial y} = \frac{V_L}{(dx)^2} \frac{\partial C}{\partial t} \tag{8.54}$$

and the momentum equations as

$$\frac{DU}{Dt} = -c_p^2 \frac{\partial p}{\partial x} + \varepsilon_d \left(\frac{\partial^2 U}{\partial x^2} + \frac{\partial^2 U}{\partial y^2} \right) \tag{8.55}$$

$$\frac{DV}{Dt} = -c_p^2 \frac{\partial p}{\partial y} + \varepsilon_d \left(\frac{\partial^2 V}{\partial x^2} + \frac{\partial^2 V}{\partial y^2} \right) + \frac{F_b}{\rho(dx)^2} \tag{8.56}$$

where p is the non-dimensional pressure, U and V are the local fluid velocities, c_p is a coefficient related to the pressure (elastic) waves celerity, ε_d is the eddy viscosity, C is the particle concentration (number of particles in a mesh of area dx^2), V_L is the volume of air corresponding to the particles entering at each time step and F_b is the per unit mass vertical buoyant driving force induced by the air bubbles (Koutitas and Gousidou-Koutita 2004). Incorporation of the notion of fluid pseudo-compressibility facilitates the solution algorithm, since the transient flow state, before reaching the final equilibrium, can be resolved in a way similar to that of free surface flows. Thus, the surrogate Lagrangian particles are transported by the local U and V velocities, diffused by the local ε_d values and always maintain a vertical velocity (U_b) in excess of the fluid velocity (V). In that case the absolute vertical velocity of the particles/bubbles is $V + U_b$.

Furthermore two of the remaining challenges are

1. To estimate the drag on each bubble and the vertical steady-state buoyancy velocity
2. To correlate the drag force on each bubble to the force exercised by the surrogate particle used in the simulation procedure to represent a number of air bubbles and to correlate such a particle with a gas volume

The drag on a bubble is given by the formula

$$F_d = \frac{1}{2}C_d\rho U_b^2\pi R_b^2 \tag{8.57}$$

where C_d is the drag coefficient related to the local Reynolds number

$R_e = \dfrac{U_b D_b}{\nu}$ by the relation

$$C_d = 0.4 + \frac{24}{R_e} + \frac{6}{1+\sqrt{R_e}} \tag{8.58}$$

From the equilibrium of buoyancy force and the drag force on the bubble, the final U_b velocity is given by

$$U_b = \sqrt{\frac{8gR_b}{3C_d}} \tag{8.59}$$

Equations 8.57 to 8.59 can only be solved iteratively.

Given the two main air-bubble flow parameters, q_g and R_b, the number of bubbles, n_b, entering the flow field is estimated as

$$n_b = \frac{q_g}{\frac{4}{3}\pi R_b^3} \tag{8.60}$$

By assuming a large total number of surrogate particles, i_p, the number of particles, n_p, entering the solution domain per unit time (second) is

$$n_p = \frac{i_p}{n_t \Delta t} \tag{8.61}$$

where n_t is the total number of time steps. Then, the equivalent drag force divided by the density, acting on each particle, is

$$T_p = \frac{n_b F_d}{n_p \rho} \tag{8.62}$$

The equivalent gas volume of the surrogate particles entering at each time step is

$$V_L = \frac{q_g}{\dfrac{i_p}{n_t}} \tag{8.63}$$

Finally, the last term in Equation 8.56 is correlated to the number of particles as

$$\frac{F_b}{\rho (dx)^2} = T_p \frac{C}{(dx)^2} \tag{8.64}$$

Equations 8.54 to 8.64 can be used to effectively simulate the air-bubble driven flow by means of surrogate Lagrangian particles.

Example 8.9

The following application investigates the flow circulation pattern in a two-dimensional domain, induced by the movement of rising air bubbles. In order to accommodate for volume characteristics in a two-dimensional vertical domain, a unit-width in the third direction has been used. Initially, the gas flow and bubble characteristics were established followed by the introduction of the surrogate Lagrangian particles. The solution domain is 30 m × 30 m discretized into equal segments of $\Delta x = \Delta y = 1$ m and the time step is $\Delta t = 0.05$ s. The data used for the simulation were either given or calculated as follows:

Gas flow = 0.001 m³/s
Bubble diameter = 0.002 m
Number of air bubbles (Equation 8.60) = 238,850 bubbles/m/s
Bubble final velocity (trial-and-error equations, Equations 8.57 to 8.59) = 0.35 m/s
Bubble drag force (trial-and-error equations, Equations 8.57 to 8.59) = 1.27×10^{-4} N
Total number of surrogate particles = 8000
Total number of time steps = 4000
Number of particles introduced per unit time (Equation 8.61) = 40 particles/m/s
Equivalent drag force acting on each particle (Equation 8.62) = 7.55×10^{-4} m³/s²
Equivalent gas volume of the particles per time (Equation 8.63) = 5×10^{-4} m²/particle

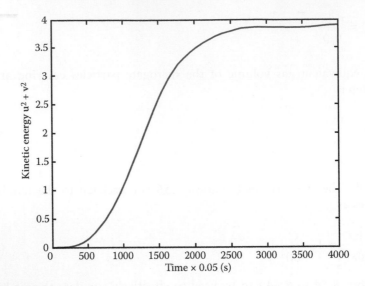

Figure 8.14 Stabilization of the kinetic energy from cold start to quasi-steady-state conditions.

The boundary conditions applied in the case of bubble-generated flow are as follows:

On the particles-source point a number of particles equal to $n_p\Delta t$ is introduced in each time step.

On the lateral boundaries the U velocity is related to the excess pressure head (Δp) as $U = \pm\Delta pc_p$.

On the upper and lower boundaries a zero-flow condition is applied ($V = 0$).

The surrogate particles reaching the boundaries of the flow domain are eliminated and are not included in the subsequent dynamic computations.

Other data utilized for the simulation were

Pressure waves celerity = 4 m/s
Eddy viscosity = 0.1 m²/s
Coordinates of the particle point source: $x_o = 14.5$ m; $y_o = 1.5$ m

The simulation was conducted for a total number of 4000 time steps. Since the simulation started from a cold start (p = U = V = 0), it was necessary that the solution reached quasi-steady-state conditions. Figure 8.14 shows that the kinetic energy was stabilized after 3000 time steps. A symmetric two-cell quasi-steady-state flow pattern and a time-frozen particle concentration distribution are illustrated, respectively, in Figures 8.15 and 8.16. The stochastic path followed by the particles is clearly depicted in Figure 8.16.

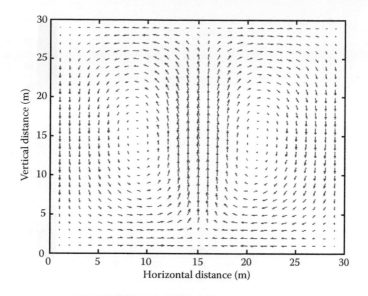

Figure 8.15 Quasi-steady-state velocities field.

Computer code 8.9

```
% Example 8.9 Air Bubbles Induced Flow
% qg = Air discharge [m^3/s];
% wb = Bubbles/particle rising velocity [m/s];
% cp = Pseudo compressibility coefficient [m/s];
% xo, yo = Particles origin coordinates [m];
% tau = Drag force divided by the density per particle [m^3/
s^2];
% ed = Eddy diffusion coefficient [m^2/s];
% ip = Total number of particles;
% Dd = Spatial step same in both directions [m];
% Dt = Time step [s];
% nx = Number of steps in the x-direction;
% ny = Number of steps in the y-direction;
% nt = Number of time steps;
clc; clear all; close all;
% Input data;
qg=0.001;
wb=0.35;
cp=4;
xo=14.5;
yo=1.5;
tau=0.000755;
ed=0.1;
ip=8000;
Dd=1;
```

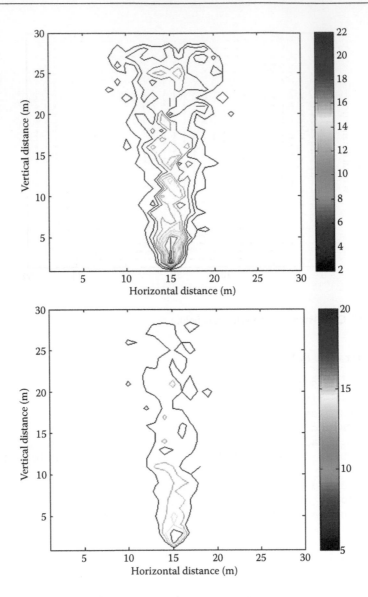

Figure 8.16 Distribution of discrete particles concentrations.

```
Dt=0.05;
nx=30;
ny=30;
nt=4000;
% Initialization of variables;
for j=1:ny
    for i=1:nx
        h(i,j)=1;
```

```
            hn(i,j)=1;
            u(i,j)=0;
            un(i,j)=0;
            v(i,j)=0;
            vn(i,j)=0;
            c(i,j)=0;
            co(i,j)=0;
        end
end
uu=0;
vv=0;
% Number of particles entering the field per unit time;
ipp=fix(ip/nt);
vol=qg*Dt/ipp;
for i=1:ipp
    x(i)=xo;
    y(i)=yo;
end
for k=1:nt
    ipt=ipp*k;
    if ipt>ip
        ipt=ip;
    end
    D=sqrt(6*ed/Dt);
    for i=ipt-ipp+1:ipt
        x(i)=xo;
        y(i)=yo;
    end
    % Estimation of pressure head;
    for j=2:ny-2
        for i=2:nx-2
            hn(i,j)=h(i,j)-Dt/Dd/2*(u(i+1,j)-u(i,j)+v(i,j+1)-
            v(i,j))+(c(i,j)-co(i,j))*vol/Dd^2;
        end
    end
    for j=1:ny-1
        hn(1,j)=hn(2,j);
        hn(nx-1,j)=hn(nx-2,j);
    end
    for i=1:nx-1
        hn(i,1)=hn(i,2);
        hn(i,ny-1)=hn(i,ny-2);
    end
    for j=1:ny-1
        for i=1:nx-1
            h(i,j)=hn(i,j);
            co(i,j)=c(i,j);
        end
    end
    % Computation of velocities along the x-axis;
```

```
for j=2:ny-2
    for i=3:nx-2
        un(i,j)=u(i,j)-Dt/Dd*cp^2*(h(i,j)-h(i-1,j))+ed*5
        *Dt/Dd^2*(u(i+1,j)+u(i-1,j)+u(i,j+1)+u(i,j-1)-
        4*u(i,j));
    end
end
for j=1:ny
    un(2,j)=-(h(2,j)-1)*cp/h(2,j);
    un(nx-1,j)=(h(nx-2,j)-1)*cp/h(nx-2,j);
end
for i=1:nx
    un(i,1)=un(i,2);
    un(i,ny-1)=un(i,ny-2);
end
% Computation of velocities along the y-axis;
for j=3:ny-2
    for i=2:nx-2
        vn(i,j)=v(i,j)-Dt/Dd*cp^2*(h(i,j)-h(i,j-
1))+ed*5*Dt/Dd^2*(v(i,j+1)+v(i,j-1)+v(i-1,j)+v(i+1,j)-
4*v(i,j))+Dt*(c(i,j)+c(i,j-1))/2/Dd^2*tau;
    end
end
for j=1:ny
    vn(1,j)=vn(2,j);
    vn(nx-1,j)=vn(nx-2,j);
end
% Updating of the velocity values;
for i=1:nx
    for j=1:ny
        u(i,j)=un(i,j);
        v(i,j)=vn(i,j);
    end
end
% Particle relocation by advection, buoyancy and
diffusion;
for i=1:ipt
    X=fix(x(i)/Dd)+1;
    Y=fix(y(i)/Dd)+1;
    if X<nx-1 && X>1 && Y<ny-1 && Y>1
        uu=u(X,Y)+(u(X+1,Y)-u(X,Y))*(x(i)-(X-1)*Dd)/Dd;
        vv=v(X,Y)+(v(X,Y+1)-v(X,Y))*(y(i)-(Y-1)*Dd)/Dd;
    end
    tempx=x(i);
    tempy=y(i);
    ax=rand;
    x(i)=x(i)+Dt*(uu+(2*ax-1)*D);

    ay=rand;
```

```
        y(i)=y(i)+Dt*(vv+(2*ay-1)*D+wb);
    end
    % Estimation of particle concentration in each cell;
    for j=1:ny-1
        for i=1:nx-1
            c(i,j)=0;
        end
    end
    for i=1:ipt
        X=fix(x(i)/Dd)+1;
        Y=fix(y(i)/Dd)+1;
        if X<nx-1 && X>1 && Y<ny-1 && Y>1
            c(X,Y)=c(X,Y)+1;
        end
    end
    % Estimation of the kinetic energy;
    KE=0;
    for i=2:nx-2
        for j=2:ny-2
            KE=KE+u(i,j)^2+v(i,j)^2;
        end
    end
    KinE(k)=KE;
    index=k
end
plot(1:k,KinE,'Linewidth',1.5);
xlabel('Time x 0.05 [s]')
ylabel('Kinetic energy u^2+v^2')
figure
[i,j]=meshgrid(1:1:nx,1:1:ny);
quiver(i,j,u',v');
xlabel('Horizontal distance [m]')
ylabel('Vertical distance [m]')
figure
contour(i,j,c')
xlabel('Horizontal distance [m]')
ylabel('Vertical distance [m]')
```

PROBLEM 8.9

Solve the same problem by making the suggested modifications while keeping the rest of the data constant:

1. Change the air pseudo-compressibility constant from 4 m/s to 1 and 8 m/s.
2. Change the eddy viscosity from 0.1 m²/s to 0.005, 0.05 and 0.2 m²/s.

3. Change the location of the air-bubble point source from $x_o = 14.5$ m, $y_o = 1.5$ m to $x_o = 4.5$ m, $y_o = 1.5$ m.
4. Change the bubble diameter from 0.002 m to 0.001 m.
5. Use two air-bubble point sources of same characteristics located at $x_o = 7.5$ m, $y_o = 1.5$ m and $x_o = 22.5$ m, $y_o = 1.5$).

Run the simulations, compare the results and comment on the behaviour of the induced flow circulation and the plume of rising bubbles.

Chapter 9

Other numerical methods

9.1 OVERVIEW

In addition to the finite differences (FD) method, there are other numerical approaches that have been developed and widely applied for the solution of differential equations. Overall, these alternate approaches have a domain discretization ability superior to the finite differences method but lack its simplicity. In the following, a brief introduction will be provided of the more well-known alternate numerical methods – the weighted residual method (WRM), the finite elements method (FEM) and the boundary elements method (BEM).

9.2 WEIGHTED RESIDUAL METHODS

The weighted residual method was developed as an approximation technique for solving differential equations and served as a prelude to the finite elements method. The basic concept of the WRM can be described as follows. Consider a linear differential operator L acting on a function $f(x)$ to produce a function $g(x)$:

$$L[f(x)] = g(x) \tag{9.1}$$

Let the unknown function $f(x)$ be approximated by a finite series,

$$f \approx \hat{f} = \sum_{i=1}^{n} \alpha_i \varphi_i \tag{9.2}$$

where φ_i are 'trial' or 'basis' functions selected from a linearly independent set and α_i are unknown constants. Substitution of the approximated function (Equation 9.2) into the differential operator (Equation 9.1) results in an error:

$$R(x) = L[\hat{f}(x)] - g(x) \neq 0 \tag{9.3}$$

where $R(x)$ is the error or residual. The main concept of the WRM is to minimize the error by forcing the residual to zero in some average sense over the solution domain D. Thus,

$$\int_D R(x)w_i \, dx = 0 \tag{9.4}$$

where w_i are the weight functions. This procedure results in a set of n-number of algebraic equations for the n-number of α_i coefficients. During the selection of the trial functions, special attention should be given ensuring they satisfy the boundary conditions. Depending on the choice of the weight functions, there are different variations of the WRM, as discussed in the following sections.

9.2.1 Collocation method

In the collocation method the weight function is the Dirac function $\delta(x)$ defined as

$$\delta(x) = 0 \text{ for } x < x_i - \varepsilon \text{ and } x_i + \varepsilon < x, \text{ while } \int_{x_i-\varepsilon}^{x_i+\varepsilon} \delta(x_i) \, dx = 1 \tag{9.5}$$

where ε is a very small positive number. Using Equation 9.5 as the weight function, integration of Equation 9.4 leads to

$$\int_D R(x)\delta(x_i) \, dx = R(x_i) = 0 \tag{9.6}$$

Thus, in the collocation method the unknown constants, α_i, are calculated by setting the residuals equal to zero on a selected number of points within the interior of the solution domain. The number of points is equal to the number of the unknown coefficients.

9.2.2 Sub-domain method

In the sub-domain method, instead of forcing the residual to vanish at selected points, the solution domain is divided into sub-domains D_i and then the residual is forced to vanish in an average sense over each of the sub-domains. The number of sub-domains is selected as equal to the number of the unknown coefficients. Thus the relation reads

$$\frac{1}{D_i} \int_{D_i} R(x)\delta^*(D_i) \, dx = \frac{1}{D_i} \int_{D_i} R(x) \, dx = 0 \tag{9.7}$$

where $\delta^*(D_i)$ is a modified Dirac function which applies over the sub-domain D_i rather than at a point.

9.2.3 Least squares method

In the least squares method, the residual is minimized in a least squares sense. For that purpose, let

$$P = \int_D R(x)R(x)\,dx = \int_D R^2(x)\,dx \tag{9.8}$$

Then the residual is minimized by setting the partial derivatives of the residual, with respect to the unknown constants, equal to zero:

$$\frac{\partial P}{\partial \alpha_i} = 2\int_D R(x)\frac{\partial R}{\partial \alpha_i}\,dx = 0 \tag{9.9}$$

Therefore, the weight functions are the derivatives of the residuals with respect to the unknown constants of the trial function.

9.2.4 Method of moments

In the method of moments, the weight functions are selected from a family of power polynomials as

$$w_i = x^i, \text{ where } i = 0, 1, 2, 3, \ldots \tag{9.10}$$

and

$$\int_D R(x)x^i\,dx = 0 \tag{9.11}$$

The number of the selected power terms (i) equals the number of the unknown constants (α_i).

9.2.5 Galerkin method

In the Galerkin method, the weight function is defined as the derivative of the trial function with respect to the unknown constants:

$$\int_D R(x)w_i\,dx = \int_D R(x)\frac{\partial \hat{f}}{\partial \alpha_i}\,dx = \int_D R(x)\varphi_i(x)\,dx = 0 \tag{9.12}$$

where the trial functions are the same as those defined in Equation 9.2.

Normally, the accuracy of the approximated solution should improve by increasing the order of the trial functions in Equation 9.2. However, an increase in the number of the trial functions causes substantial complexity on the required preliminary steps for the weighted residual methods.

Example 9.1

The weighted residuals method is demonstrated by using the ordinary differential equation for a damped harmonic oscillator written as

$$m\frac{d^2f}{dx^2} + c\frac{df}{dx} + kx = 0 \qquad (9.13)$$

where m, c and k are constants. The exact solution of Equation 9.13 is

$$f = \exp\left[\left(\frac{-c \pm \sqrt{c^2 - 4mk}}{2m}\right)x\right] \qquad (9.14)$$

Depending on the discriminant $\Delta = c^2 - 4mk$, the Equation 9.14 can have two real ($\Delta > 0$), one real ($\Delta = 0$) or two complex solutions ($\Delta < 0$). Thus the general solution reads

$$f = c_1 e^{\lambda_1 x} + c_2 e^{\lambda_2 x} \qquad (9.15)$$

where c_1 and c_2 are constants depending on the boundary conditions.
By assuming m = 2, c = −6 and k = 4, the two exponent coefficients are $\lambda_1 = 1$ and $\lambda_2 = 2$, and the solution becomes

$$f = c_1 e^x + c_2 e^{2x} \qquad (9.16)$$

The two terms in the solution are linearly independent since the Wronskian is different than zero:

$$W(e^x, e^{2x}) = \begin{bmatrix} e^x & e^{2x} \\ e^x & 2e^{2x} \end{bmatrix} \neq 0 \qquad (9.17)$$

Furthermore, if the solution domain is $0 \le x \le 1$ and the boundary conditions are $f(x = 0) = 1$ and $f(x = 1) = 0$, then $c_1 = 1.5819767$ and $c_2 = -0.5819767$. Therefore, under the aforementioned conditions, the exact (analytical) solution is

$$f = 1.5819767e^x - 0.5819767e^{2x} \qquad (9.18)$$

Then, in order to apply the weighted residual method, a trial function is selected as

$$\hat{f} = \alpha_0 + \alpha_1 x + \alpha_2 x^2 \tag{9.19}$$

Since the trial function needs to satisfy the aforementioned boundary conditions, the unknown constants are estimated as $\alpha_0 = 1$ and $\alpha_2 = 1 - \alpha_1$, which leads to

$$\hat{f} = 1 + \alpha_1 x - (1 + \alpha_1) x^2 = 1 - x^2 + x(1-x)\alpha_1 \tag{9.20}$$

Substitution of Equation 9.20 into Equation 9.13 results in the residual as follows:

$$R = -2m + (c - 2m - 2cx)\alpha_1 + kx - 2cx$$
$$= (12x - 10)\alpha_1 + 16x - 4 \neq 0 \tag{9.21}$$

Collocation method: The collocation method requires the selection of one point within the solution domain $0 \le x \le 1$ since there is only one unknown (α_1). After selecting the point $x_1 = 0.5$, then the residual is set equal to zero according to Equations 9.6 and 9.21:

$$R = (12 \cdot 0.5 - 10)\alpha_1 + 16 \cdot 0.5 - 4 = -4\alpha_1 + 4 = 0 \tag{9.22}$$

From Equation 9.22 $\alpha_1 = 1$ and the approximate solution is

$$\hat{f} = 1 + x - 2x^2 \tag{9.23}$$

Sub-domain method: For the sub-domain method, the solution domain is discretized into a single sub-domain D (0 to 1). In this method the solution is obtained by solving the following equation resulting from Equation 9.7:

$$R = \int_0^1 [-2m + (c - 2m - 2cx)\alpha_1 + kx - 2cx] dx$$

$$\tag{9.24}$$

$$= \int_0^1 [(12x - 10)\alpha_1 + 16x - 4] dx = 0$$

After integration, the constant parameter is found as $\alpha_1 = 1$, so that the approximate solution is the same as the one obtained by the collocation method (Equation 9.23).

Least squares method: For the least squares method, first, the derivative of the residual with respect to the unknown constant is calculated as

$$\frac{\partial R}{\partial \alpha_1} = c - 2m - 2cx = 12x - 10 \tag{9.25}$$

Then, based on Equation 9.9, the following equation is obtained for the solution of the unknown constant:

$$\int_0^1 R \frac{\partial R}{\partial \alpha_1} dx = \int_0^1 [-2m + (c - 2m - 2cx)\alpha_1 + kx - 2cx](c - 2m - 2cx) dx$$

(9.26)

By substituting the constants m, c and k, and after some arithmetic manipulation, Equation 9.26 becomes

$$\int_0^1 R \frac{\partial R}{\partial \alpha_1} dx = \int_0^1 [144\alpha_1 x^2 + 192x^2 - 240\alpha_1 x - 208x + 40] dx = -72\alpha_1 = 0$$

(9.27)

Therefore, $\alpha_1 = 0$ and the approximate solution is given as

$$\hat{f} = 1 - x^2$$

(9.28)

Method of moments: In the method of moments, the weight function selected is $w_0 = x^0 = 1$. Then, according to Equation 9.11, the method becomes identical to the sub-domain method (Equation 9.24) and the approximate solution is the same as in Equation 9.23.

Galerkin method: Finally, for the Galerkin method the weight function specific to this example is

$$w = \frac{\partial \hat{f}}{\partial \alpha_1} = x(1 - x)$$

(9.29)

and the equation for the unknown constant α_1 is derived from the integral:

$$\int_0^1 Rw \, dx = \int_0^1 R \frac{\partial \hat{f}}{\partial \alpha_1} dx = 0$$

(9.30)

After substitution of the constants m, c and k, and after some arithmetic manipulation, Equation 9.30 yields

$$\int_0^1 R \frac{\partial \hat{f}}{\partial \alpha_1} dx = \int_0^1 [-(12\alpha_1 + 16)x^3 + (22\alpha_1 + 20)x^2 - (10\alpha_1 - 4)x] dx = -2\alpha_1 + 2 = 0$$

(9.31)

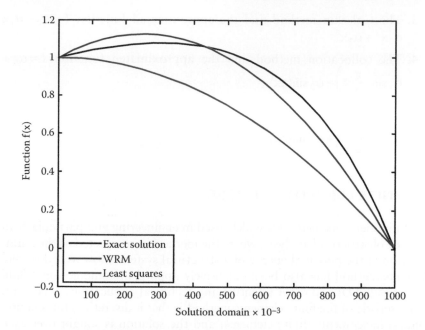

Figure 9.1 Comparison of the weighted residual methods.

From Equation 9.31 it can be seen that the approximate equation is again the same as in Equation 9.23. Therefore, the four methods (collocation, sub-domain, moments and Galerkin) result in the same approximated solution (Equation 9.23), while the method of least squares gives another approximating solution (Equation 9.28). Comparison of the results shows overall that the methods are not very accurate and this is particularly true for the least squares method (Figure 9.1). This inaccuracy occurs because the selected weight functions are incompatible with the physical problem.

PROBLEM 9.1

Solve Equation 9.13 for the same constants m, c and k and the same boundary conditions by using:

1. The moments method and the approximated function $\hat{f} = \alpha_0 + \alpha_1 x + \alpha_2 x^2 + \alpha_3 x^3$

2. The least squares method and the approximated function $\hat{f} = \alpha_0 + \alpha_1 x + \alpha_2 x^2 + \alpha_3 x^3$

3. The Galerkin method and the approximated function $\hat{f} = \alpha_0 + \alpha_1 x$ $\alpha_2 x^2 + \alpha_3 x^3$

4. The collocation method and the approximated function $\hat{f} = \alpha_0 + \alpha_1 \sin\left(\frac{\pi}{2}x\right) + \alpha_2 \sin(\pi x)$

5. The Galerkin method and the approximated function $\hat{f} = \alpha_0 + \alpha_1 \sin\left(\frac{\pi}{2}x\right) + \alpha_2 \sin(\pi x)$

9.3 FINITE ELEMENTS METHOD

The finite elements method is widely used in engineering and particularly in the area of structural analysis, where the method is interpreted as the maximization of the potential energy of a structural system. However, the finite elements method has also been extensively applied in computational fluid dynamics (CFD) and other scientific and technical applications. The main characteristic of the finite elements method is that it discretizes the solution domain in segments (finite elements) and the solution is sought over each element. The elements are connected to one another through their common nodes. The resulting algebraic difference equations are similar to the ones derived by the finite differences method but with some weighted factors. The advantage of the finite element method is its ability to easily handle even the most complex boundary shapes. The disadvantage of the method is the somewhat laborious pre-treatment of the differential equations before they are converted into difference equations. For that purpose, there are two main approaches: the Rayleigh-Ritz method and the Galerkin method (that has already been discussed as one of the weighted residual methods).

9.3.1 Rayleigh-Ritz method

The Rayleigh-Ritz method can be explained in terms of the theory of variational principles (Gelfand and Fomin 1963). In general, variational principles investigate the one-to-one correspondence and equivalence between certain differential equations and their functionals. For those values of the functions, f, which are solutions of the original differential equation (known as the Euler's equation), the functionals I[f] become stationary (reaching a relative maximum or minimum value).

For example, for a second-order differential equation, a functional I[f] is defined as

$$I[f] = \int_D F(x, f, f', f'') \, dx \tag{9.32}$$

where f is an unknown function and the prime denotes the order of differentiation. Then the relationship, whenever it exists, between the functional and its Euler equation is given as

$$I[f] = \int_D F(x, f, f', f'') \, dx = \text{stationary} \Leftrightarrow F_f - \frac{d}{dx}(F_{f'}) + \frac{d^2}{dx^2}(F_{f''}) = 0 \quad (9.33)$$

where $F_{f^n} = \dfrac{\partial F}{\partial f^n}$ and n is the order of differentiation.

Example 9.2

Let us consider the functional

$$I[f] = \int_D F(f, f') \, dx = \int_D \left[\frac{1}{2}\left(\frac{df}{dx}\right)^2 + gf \right] dx \quad (9.34)$$

where g is a known function of x. The function that makes the functional stationary must satisfy the Euler equation

$$\frac{d^2 f}{dx^2} - g(x) = 0 \quad (9.35)$$

Let $g(x) = -x$ and the boundary conditions for Equation 9.35 are taken as $f(0) = f(1) = 0$, $(0 \le x \le 1)$. An approximate solution of the problem can be accomplished by assuming a third-order power series function. Then, the approximate function that also satisfies the boundary conditions reads

$$\hat{f} = x(x - 1)(\alpha_1 + \alpha_2 x) \quad (9.36)$$

Substituting Equation 9.36 into Equation 9.34 and integrating from 0 to 1 yields

$$I[\alpha_1, \alpha_2] = \frac{\alpha_1^2}{6} + \frac{\alpha_2^2}{15} + \frac{\alpha_1 \alpha_2}{6} + \frac{\alpha_1}{12} + \frac{\alpha_2}{20} \quad (9.37)$$

A stationary solution of Equation 9.34 can be obtained according to the conditions

$$\frac{\partial I}{\partial \alpha_1} = \frac{2\alpha_1}{6} + \frac{\alpha_2}{6} + \frac{1}{12} = 0 \quad (9.38)$$

$$\frac{\partial I}{\partial \alpha_2} = \frac{2\alpha_2}{15} + \frac{\alpha_1}{6} + \frac{1}{20} = 0 \qquad (9.39)$$

that lead to $\alpha_1 = \alpha_2 = -\frac{1}{6}$, and to the approximate solution

$$\widehat{f} = \frac{1}{6}x(1-x^2) \qquad (9.40)$$

Notable in this particular case, the approximate solution coincides with the exact solution of the Euler equation (Equation 9.35).

PROBLEM 9.2

Use variational calculus and the Rayleigh-Ritz method for the following cases:

1. Solve the same problem as in Example 9.2 by using the approximated function: $\widehat{f} = x(x-1)(\alpha_1 + \alpha_2 x + \alpha_3 x^2)$.

2. Find the Euler's equation for the functional $I[f] = \int_D \left[\sqrt{\left(\dfrac{df}{dx}\right)^2 + 1} \right] dx$.

3. Find the solution of the functional given in question (2) in the domain $x_A \le x \le x_B$ under the boundary conditions $f(x = x_A) = f_A$ and $f(x = x_B) = f_B$.

4. Find the Euler's equation of the functional $I[f] = \int_1^2 \left[\dfrac{\sqrt{1 + \left(\dfrac{df}{dx}\right)^2}}{x} \right] dx$

   under the boundary conditions $f(x = 1) = 0$ and $f(x = 2) = 1$.

5. Find the Euler's equation of the functional $I[f] = 2\pi \int_{x_A}^{x_B} \left[f\sqrt{\left(\dfrac{df}{dx}\right)^2 + 1} \right] dx$.

9.3.2 Finite elements method

In the weighted residual methods and the Rayleigh-Ritz method the approximate solutions were defined and applied for the entire solution domain. In the finite elements method the approximate functions are limited to a single segment of the domain that is the finite element. For that purpose, a shape function, also known as trial or base function, is defined for each element. For one-dimensional space and for an element (e_i), the simplest shape functions can be defined by linear expressions as

$$N_i^{(e_i)}(x) = \frac{x_{i+1} - x}{x_{i+1} - x_i} = \frac{x_{i+1} - x}{l_i^{(e_i)}} \qquad (9.41)$$

$$N_{i+1}^{(e_i)}(x) = \frac{x - x_i}{x_{i+1} - x_i} = \frac{x - x_i}{l_i^{(e_i)}} \tag{9.42}$$

where $l_i^{(e_i)}$ is the length of the finite element. The preceding shape functions and the discretization of the one-dimensional field into finite elements are demonstrated in Figure 9.2.

Once the shape functions are defined, the unknown functions are described within the element as

$$\hat{f}^{(e_i)} = N_i^{(e_i)} f_i + N_{i+1}^{(e_i)} f_{i+1} = \begin{bmatrix} N_i^{(e_i)} & N_{i+1}^{(e_i)} \end{bmatrix} \begin{Bmatrix} f_i \\ f_{i+1} \end{Bmatrix} = [N]^{(e_i)} \{f\}^{(e_i)} \tag{9.43}$$

Equation 9.43 constitutes the 'local' system. In order to obtain the 'global' system, all of the local systems (M-number of elements) need to be assembled appropriately as

$$\sum_{i=1}^{M} \hat{f}^{(e_i)} = \sum_{i=1}^{M} [N]^{(e_i)} \{f\}^{(e_i)} \tag{9.44}$$

In addition to the linear elements, quadratic elements can be applied. A quadratic element consists of three nodes (nodes 1 and 3 at the boundaries of the element and node 2 at the middle). A local coordinate system (ξ) is set for each element so that $\xi = -1$ at node 1, $\xi = 0$ at node 2 and $\xi = 1$ at node 3 (Figure 9.3).

Then the unknown function is approximated by $\hat{f}(\xi) = \sum_{i-1}^{3} N_i f_i$, where the shape functions are defined as

$$N_1 = -\frac{1}{2}\xi(1-\xi) \tag{9.45}$$

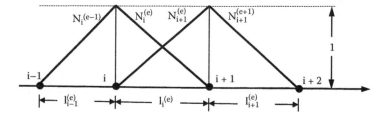

Figure 9.2 One-dimensional linear shape functions.

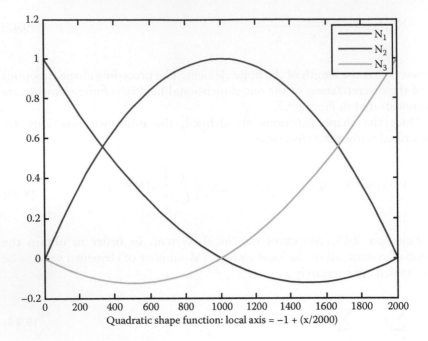

Figure 9.3 Quadratic finite elements.

$$N_2 = 1 - \xi^2 \tag{9.46}$$

$$N_3 = \frac{1}{2}\xi(1 + \xi) \tag{9.47}$$

so that $\widehat{f}(-1) = f_1$, $\widehat{f}(0) = f_2$ and $\widehat{f}(1) = f_3$.

After defining the approximated expressions for the unknown functions, the solution can be accomplished by using either the Rayleigh-Ritz or the Galerkin method. In both cases the domain is divided into a number of finite elements. Then the approximate function is substituted into the governing equation and a numerical solution is being sought for the unknown values of the function at each node.

Example 9.3

Consider the same problem as the one in Example 9.2 (Equation 9.34 or Equation 9.35) and seek the solution by applying both the Rayleigh-Ritz and the finite elements methods. For simplicity, the solution domain is discretized into two equal finite elements. Using the shape functions as defined by Equations 9.41 and 9.42 yields

$$N_1^{(1)}(x) = \frac{0.5 - x}{0.5}, N_2^{(1)}(x) = \frac{x}{0.5}, N_2^{(2)}(x) = \frac{1 - x}{0.5}, N_3^{(2)}(x) = \frac{x - 0.5}{0.5} \qquad (9.48)$$

and the derivatives are estimated as

$$\frac{dN_1^{(1)}}{dx} = -2, \frac{dN_2^{(1)}}{dx} = 2, \frac{dN_2^{(2)}}{dx} = -2, \frac{dN_3^{(2)}}{dx} = 2 \qquad (9.49)$$

Since during the integration process the product of the shape functions raised to different powers appears very often, a useful formula that facilitates the calculations is

$$\int_{l^{(e)}} N_i^\alpha N_{i+1}^\beta dx = \frac{\alpha\beta \, l^{(e)}}{(\alpha + \beta + 1)} \qquad (9.50)$$

Rayleigh-Ritz method: First, the solution is being sought by using the Rayleigh-Ritz method. Thus the approximate solution (Equation 9.43) is substituted into the functional (Equation 9.34) leading to

$$I[f] = \int_0^1 \left[\frac{1}{2} \left(\frac{df}{dx} \right)^2 - xf \right] dx \qquad (9.51)$$

or

$$I[f] = \int_0^{0.5} \left[\frac{1}{2} \left(\frac{d}{dx} \begin{bmatrix} N_1^{(1)} & N_2^{(1)} \end{bmatrix} \begin{Bmatrix} f_1 \\ f_2 \end{Bmatrix} \right)^2 \right] dx - \int_0^{0.5} \left[x \begin{bmatrix} N_1^{(1)} & N_2^{(1)} \end{bmatrix} \begin{Bmatrix} f_1 \\ f_2 \end{Bmatrix} \right] dx$$
$$+ \int_{0.5}^1 \left[\frac{1}{2} \left(\frac{d}{dx} \begin{bmatrix} N_2^{(2)} & N_3^{(2)} \end{bmatrix} \begin{Bmatrix} f_2 \\ f_3 \end{Bmatrix} \right)^2 \right] dx - \int_0^{0.5} \left[x \begin{bmatrix} N_2^{(2)} & N_3^{(2)} \end{bmatrix} \begin{Bmatrix} f_2 \\ f_3 \end{Bmatrix} \right] dx \qquad (9.52)$$

Accounting for the boundary conditions $f_1 = f_3 = 0$ and after integrating, Equation 9.52 is reduced to

$$I[f_2] = 2f_2^2 - 0.25f_2 \qquad (9.53)$$

Then for the functional to become stationary

$$\frac{dI[f_2]}{df_2} = 4f_2 - 0.25 = 0 \Rightarrow f_2 = 0.625 \tag{9.54}$$

Therefore, the value of the function at x = 0.5 (node 2) is equal to 0.625, and that coincides with the exact solution (Equation 9.40).

Galerkin method: Now the solution is being sought by using the Galerkin method. For that purpose, Euler's equation (Equation 9.35) is used and again the solution domain is divided into two equal-length finite elements. However, since the shape function is linear and Euler's equation is of the second-order, the second derivative of the trial function becomes zero. To avoid this difficulty, the theorem of integration by parts is applied (Chung 1978):

$$\int_{x_1}^{x_2} \frac{du}{dx} v \, dx = uv \Big|_{x_1}^{x_2} - \int_{x_1}^{x_2} u \frac{dv}{dx} dx \tag{9.55}$$

where

$$u = \frac{df}{dx}, \quad v = N \tag{9.56}$$

After integrating by parts, the Galerkin method yields

$$\int_0^1 \{L[\hat{f}(x)] - g(x)\} N_2 dx =$$

$$-\int_0^{0.5}\left(\frac{d}{dx}\left[N_1^{(1)} \quad N_2^{(1)}\right]\left\{\begin{matrix}f_1\\f_2\end{matrix}\right\}\right)\left(\frac{dN_2^{(1)}}{dx}\right)dx + \int_0^{0.5} xN_2^{(1)} dx + \frac{d}{dx}\left[N_1^{(1)} \quad N_2^{(1)}\right]\left\{\begin{matrix}f_1\\f_2\end{matrix}\right\}N_2^{(1)}\Big|_0^{0.5}$$

$$-\int_{0.5}^1\left(\frac{d}{dx}\left[N_2^{(2)} \quad N_3^{(2)}\right]\left\{\begin{matrix}f_2\\f_3\end{matrix}\right\}\right)\left(\frac{dN_2^{(2)}}{dx}\right)dx + \int_{0.5}^1 xN_2^{(2)} dx + \frac{d}{dx}\left[N_2^{(2)} \quad N_3^{(2)}\right]\left\{\begin{matrix}f_2\\f_3\end{matrix}\right\}N_2^{(1)}\Big|_{0.5}^1 = 0 \tag{9.57}$$

Substituting the shape functions, accounting for the boundary conditions $f_1 = f_3 = 0$, and integrating, Equation 9.57 results in $f_2 = \frac{1}{16} = 0.625$. The solution is identical to the one obtained by the Rayleigh-Ritz method and the closed form solution.

In order to generalize the finite element Galerkin method, the local system for an element (e) of length l_e between the nodes x_i and x_{i+1} is written as

$$
\begin{bmatrix} \alpha_{i,i} & \alpha_{i,i+1} \\ \alpha_{i+1,i} & \alpha_{i+1,i+1} \end{bmatrix} \begin{Bmatrix} f_i \\ f_{i+1} \end{Bmatrix} = \begin{Bmatrix} c_i \\ c_{i+1} \end{Bmatrix}
\tag{9.58}
$$

where the matrix coefficients α and the constants c are estimated as follows:

$$
\alpha_{i,i} = \int_{(e)} \frac{dN_i}{dx} \frac{dN_i}{dx} dx = \frac{1}{l_e}
\tag{9.59}
$$

$$
\alpha_{i,i+1} = \int_{(e)} \frac{dN_{i+1}}{dx} \frac{dN_i}{dx} dx = -\frac{1}{l_e}
\tag{9.60}
$$

$$
\alpha_{i+1,i} = \int_{(e)} \frac{dN_i}{dx} \frac{dN_{i+1}}{dx} dx = -\frac{1}{l_e}
\tag{9.61}
$$

$$
\alpha_{i+1,i+1} = \int_{(e)} \frac{dN_{i+1}}{dx} \frac{dN_{i+1}}{dx} dx = \frac{1}{l_e}
\tag{9.62}
$$

$$
c_i = \int_{(e)} xN_i \, dx = \int_{(e)} x\left(\frac{x_{i+1} - x}{l_e} \right) dx = \frac{x_{i+1}\left(x_{i+1}^2 - x_i^2\right)}{2l_e} - \frac{\left(x_{i+1}^3 - x_i^3\right)}{3l_e}
\tag{9.63}
$$

$$
c_{i+1} = \int_{(e)} xN_{i+1} \, dx = \int_{(e)} x\left(\frac{x - x_i}{l_e} \right) dx = \frac{\left(x_{i+1}^3 - x_i^3\right)}{3l_e} - \frac{x_i\left(x_{i+1}^2 - x_i^2\right)}{2l_e}
\tag{9.64}
$$

By considering 10 finite elements the local contributions can be assembled into the global system that would provide the solution of Equation 9.35. As it can be seen from Figure 9.4, the finite element Galerkin solution is almost indistinguishable from the exact solution.

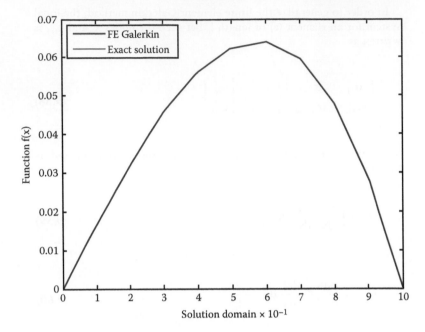

Figure 9.4 Comparison of the finite element Galerkin approximation versus the exact solution.

Computer code 9.1

```
% Example 9.1 Finite Element Method
% dx = Length of the finite element;
% imax = Number of nodes;
% a = Matrix coefficients;
% c = Vector of constants;
% f = value of functions at the nodes;
dx = 0.1;
imax = 11;
imax1 = imax-1;
for i = 1:imax
    f(i) = 0;
    c(i) = 0;
    for j = 1:imax
        a(i,j) = 0;
    end
end
for i = 1:imax1
    a(i,i)   = a(i,i)+1/dx;
    a(i,i+1) = a(i,i+1)-1/dx;
    a(i+1,i) = a(i+1,i)-1/dx;
    a(i+1,i+1) = a(i+1,i+1)+1/dx;
    c(i)   = c(i)+dx^2*(i*(i^2-(i-1)^2)/2-(i^3-(i-1)^3)/3);
    c(i+1) = c(i+1)+dx^2*((i^3-(i-1)^3)/3-(i-1)*(i^2-(i-1)^2)/2);
```

```
end
iter = 0;
for k = 1:1000
    iter = iter+1;
    difmax = 0;
    for i = 2:imax1
        f(i) = (c(i)-a(i,i+1)*f(i+1)-a(i,i-1)*f(i-1))/a(i,i);
    end
end
x = 0;
for i = 1:imax
    fexact(i) = x*(1-x^2)/6;
    x = x+dx;
end
plot(0:imax-1,f,'b','Linewidth',2)
hold on
plot(0:imax-1,fexact,'r')
xlabel('Solution domain x 10^-^1')
ylabel('Function f(x)')
legend('FE Galerkin','Exact Solution',2)
```

PROBLEM 9.3

Solve the same problem (Equation 9.34 or Equation 9.35) according to the following instructions:

1. Subdivide the solution domain into three finite elements of equal length and apply the Rayleigh-Ritz method. Calculate the solution by hand.
2. Subdivide the solution domain into three finite elements of equal length and apply the finite elements Galerkin method. Calculate the solution by hand.
3. Use the computer to solve the problem described in question (2).
4. Use the quadratic element (Equations 9.45 through 9.47) to solve the problem described in Example 9.3 utilizing one finite element Rayleigh-Ritz method.
5. Use the quadratic element (Equations 9.45 through 9.47) to solve the problem described in Example 9.3 utilizing the finite element Galerkin method.

9.3.3 Application of the finite elements method in water resources

In the previous sections, the finite elements method was demonstrated by using simple ordinary differential equations. In the following, without elaborating into much detail, the finite element Galerkin method is applied for a second-order hyperbolic partial differential equation (telegrapher's equation) and for a hyperbolic system of differential equations (Saint-Venant equations). In both cases, the resulting discretized algebraic equations are provided.

9.3.3.1 Telegrapher's equation

For this case, the telegrapher's equation (Equation 6.23) is written as

$$\frac{\partial^2 \zeta}{\partial t^2} - g \frac{\partial}{\partial x}\left(h \frac{\partial \zeta}{\partial x} \right) + \frac{\kappa}{h} \frac{\partial \zeta}{\partial t} = 0 \tag{9.65}$$

where h is the water depth and κ is a bottom friction coefficient. By considering the local matrix as

$$\widehat{\zeta}^{(e_i)} = \begin{bmatrix} N_i^{(e_i)} & N_{i+1}^{(e_i)} \end{bmatrix} \left\{ \begin{matrix} \zeta_i \\ \zeta_{i+1} \end{matrix} \right\} \tag{9.66}$$

after integration by parts of the second derivative and substitution of the approximate function, Equation 9.65 becomes

$$\sum_{e=1}^{M} \left(\int_{(e)} [N]^{(e)} \left\{ \frac{\partial^2 \zeta}{\partial t^2} \right\} N_j^{(e)} \, dx + gh^{(e)} \int_{(e)} \frac{\partial [N]^{(e)}}{\partial x} \{\zeta\} \frac{\partial N_j^{(e)}}{\partial x} \, dx \right.$$

$$\left. + \frac{\kappa}{h^{(e)}} \int_{(e)} [N]^{(e)} \left\{ \frac{\partial \zeta}{\partial t} \right\} N_j^{(e)} \, dx \right) = 0 \tag{9.67}$$

where j = i, i + 1. Equation 9.67 contains first- and second-order derivatives with respect to time that are treated by using explicit finite differences approximation as follows:

$$\frac{\partial \zeta}{\partial t} = \frac{\zeta_i^{n+1} - \zeta_i^n}{t} \tag{9.68}$$

$$\frac{\partial^2 \zeta}{\partial t^2} = \frac{\zeta_i^{n+1} - 2\zeta_i^n + \zeta_i^{n-1}}{(\Delta t)^2} \tag{9.69}$$

Then for each time step, the resulting algebraic system in matrix form reads (Koutitas 1983)

$$
\begin{bmatrix} \alpha_{i,i} & \alpha_{i,i+1} \\ \alpha_{i+1,i} & \alpha_{i+1,i+1} \end{bmatrix} \begin{Bmatrix} \zeta_i^{n+1} \\ \zeta_{i+1}^{n+1} \end{Bmatrix} = \begin{Bmatrix} c_i\left(\zeta_i^{n-1},\zeta_i^n\right) \\ c_{i+1}\left(\zeta_{i+1}^{n-1},\zeta_{i+1}^n\right) \end{Bmatrix}
\tag{9.70}
$$

and the values of the matrix and the constant vector coefficients are given as

$$
\alpha_{i,i} = \frac{\Delta x}{3}
\tag{9.71}
$$

$$
\alpha_{i,i+1} = \frac{\Delta x}{6}
\tag{9.72}
$$

$$
\alpha_{i+1,i} = \frac{\Delta x}{6}
\tag{9.73}
$$

$$
\alpha_{i+1,i+1} = \frac{\Delta x}{3}
\tag{9.74}
$$

$$
c_i = \left(2\zeta_i^n - \zeta_i^{n-1}\right)\frac{\Delta x}{3} + \left(2\zeta_{i+1}^n - \zeta_{i+1}^{n-1}\right)\frac{\Delta x}{6} - gh^{(e)}\left(\zeta_i^n - \zeta_{i+1}^n\right)\frac{(\Delta t)^2}{\Delta x}
$$
$$
- \frac{\kappa(\Delta t)^2}{h^{(e)}}\left[\frac{\left(\zeta_i^n - \zeta_i^{n-1}\right)}{3} + \frac{\left(\zeta_{i+1}^n - \zeta_{i+1}^{n-1}\right)}{6}\right]\frac{\Delta x}{\Delta t}
\tag{9.75}
$$

$$
c_{i+1} = \left(2\zeta_i^n - \zeta_i^{n-1}\right)\frac{\Delta x}{6} + \left(2\zeta_{i+1}^n - \zeta_{i+1}^{n-1}\right)\frac{\Delta x}{3} - gh^{(e)}\left(-\zeta_i^n + \zeta_{i+1}^n\right)\frac{(\Delta t)^2}{\Delta x}
$$
$$
- \frac{\kappa(\Delta t)^2}{h^{(e)}}\left[\frac{\left(\zeta_i^n - \zeta_i^{n-1}\right)}{6} + \frac{\left(\zeta_{i+1}^n - \zeta_{i+1}^{n-1}\right)}{3}\right]\frac{\Delta x}{\Delta t}
\tag{9.76}
$$

For a tidal estuary with a closed-end boundary, the boundary conditions can be defined by a sinusoidal function at the open-sea end and by a no-flux condition at the reflecting end.

9.3.3.2 Saint-Venant system of equations

The continuity and momentum balance equations written for a shallow wave constitute the Saint-Venant system of equations. By considering an exponentially widening, rectangular shape in the downstream direction estuary, the Saint-Venant system of equations for a one-dimensional case are written as

$$\frac{\partial \zeta}{\partial t} + (h + \zeta)\frac{\partial u}{\partial x} + u\frac{\partial \zeta}{\partial x} + uS_o + u(h + \zeta)\lambda = 0 \tag{9.77}$$

$$\frac{\partial \zeta}{\partial t} + u\frac{\partial u}{\partial x} + g\frac{\partial \zeta}{\partial x} - g\left[S_o - \frac{u|u|}{C_z^2(h + \zeta)}\right] = 0 \tag{9.78}$$

The width of the estuary varies according to the expression

$$b = b_o e^{\lambda x} \tag{9.79}$$

where b_o is the width upstream at the river mouth or the closed-end boundary and λ is a positive number. The Saint-Venant system is a time-dependent, non-linear partial differential hyperbolic system of equations in terms of two variables: the depth-averaged velocity $u(x,t)$ and the disturbance of the water surface elevation $\zeta(x,t)$.

The solution to the problem was sought by using a finite elements Galerkin approach. For that purpose the solution domain was discretized into space-time finite elements by using a rectangular shape function, as shown in Figure 9.5.

The values of the unknown functions $u(x,t)$ and $\zeta(x,t)$ are approximated for each element (e) as

$$\left\{\begin{matrix} \{u\}_{4x1} \\ \{\zeta\}_{4x1} \end{matrix}\right\}^{(e)} = \left[\begin{matrix} [N]_{4x4} & [0]_{4x4} \\ [0]_{4x4} & [N]_{4x4} \end{matrix}\right]\left\{\begin{matrix} \{u\}_{4x1} \\ \{\zeta\}_{4x1} \end{matrix}\right\} \tag{9.80}$$

The general form of the serendipity-type shape functions (Zienkiewicz 1971) is

$$N_k(x,t) = \frac{1}{4}(1 + \xi_o)(1 + \eta_o), \quad k = 1,2,3,4 \tag{9.81}$$

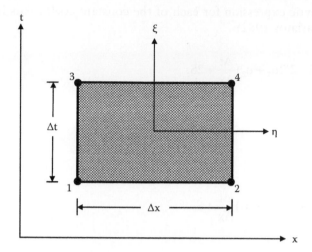

Figure 9.5 Space–time finite element.

where

$$\xi_o = \xi\xi_k = \left[\frac{2x - (x_i + x_{i-1})}{\Delta x}\right]\xi_k \tag{9.82}$$

$$\eta_o = \eta\eta_k = \left[\frac{2t - (t_{j+1} + t_j)}{\Delta t}\right]\eta_k \tag{9.83}$$

In addition, $\xi_2 = \xi_4 = \eta_3 = \eta_4 = 1$ and $\xi_1 = \xi_3 = \eta_1 = \eta_2 = -1$. This approach eliminates the need for separate treatment of time derivatives by using a finite differences scheme. Since the value of the dependent functions is known at nodes 1 and 2, either from the previous time step or the initial conditions, the number of unknowns in Equation 9.80 is reduced by half. By applying the Galerkin approach, the resulting system for each element reads

$$\begin{bmatrix} \alpha_{1,1} & \alpha_{1,2} & \alpha_{1,3} & \alpha_{1,4} \\ \alpha_{2,1} & \alpha_{2,2} & \alpha_{2,3} & \alpha_{2,4} \\ \alpha_{3,1} & \alpha_{3,2} & \alpha_{3,3} & \alpha_{3,4} \\ \alpha_{4,1} & \alpha_{4,2} & \alpha_{4,3} & \alpha_{4,4} \end{bmatrix} \begin{Bmatrix} u_3 \\ \zeta_3 \\ u_4 \\ \zeta_4 \end{Bmatrix} = \begin{Bmatrix} c_1 \\ c_2 \\ c_3 \\ c_4 \end{Bmatrix} \tag{9.84}$$

The analytic expression for each of the constant coefficients is given as follows (Scarlatos 1982):

$$\alpha_{1,1} = 6 - 2(2u_1 + u_2)\frac{\Delta t}{\Delta x} + 2S_e \qquad (9.85)$$

$$\alpha_{1,2} = -6g\frac{\Delta t}{\Delta x} \qquad (9.86)$$

$$\alpha_{1,3} = 3 + 2(2u_1 + u_2)\frac{\Delta t}{\Delta x} + S_e \qquad (9.87)$$

$$\alpha_{1,4} = 6g\frac{\Delta t}{\Delta x} \qquad (9.88)$$

$$\alpha_{2,1} = \left[-6h + (h\lambda + S_o)\Delta x + (\lambda\Delta x - 4)\zeta_1 + \left(\frac{\lambda}{4}\Delta x - 2\right)\zeta_2 \right]\frac{\Delta t}{\Delta x} \qquad (9.89)$$

$$\alpha_{2,2} = 2\left[3 - (2u_1 + u_2)\frac{\Delta t}{\Delta x} \right] \qquad (9.90)$$

$$\alpha_{2,3} = \left[6h + \frac{1}{2}(h\lambda + S_o)\Delta x + \left(\frac{\lambda}{4}\Delta x + 4\right)\zeta_1 + \left(\frac{\lambda}{4}\Delta x + 2\right)\zeta_2 \right]\frac{\Delta t}{\Delta x} \qquad (9.91)$$

$$\alpha_{2,4} = 3 + 2(2u_1 + u_2)\frac{\Delta t}{\Delta x} \qquad (9.92)$$

$$\alpha_{3,1} = 3 - 2(2u_1 + u_2)\frac{\Delta t}{\Delta x} + S_e \qquad (9.93)$$

$$\alpha_{3,2} = -6g\frac{\Delta t}{\Delta x} \qquad (9.94)$$

$$\alpha_{3,3} = 6 + 2(2u_1 + u_2)\frac{\Delta t}{\Delta x} + 2S_e \tag{9.95}$$

$$\alpha_{3,4} = 6g\frac{\Delta t}{\Delta x} \tag{9.96}$$

$$\alpha_{4,1} = \left[-6h + \frac{1}{2}(h\lambda + S_o)\Delta x + \left(\frac{\lambda}{4}\Delta x - 24\right)\zeta_1 + \left(\frac{\lambda}{4}\Delta x - 4\right)\zeta_2\right]\frac{\Delta t}{\Delta x} \tag{9.97}$$

$$\alpha_{4,2} = 3 - 2(u_1 + 2u_2)\frac{\Delta t}{\Delta x} \tag{9.98}$$

$$\alpha_{4,3} = \left[6h + (h\lambda + S_o)\Delta x + \left(\frac{\lambda}{4}\Delta x + 2\right)\zeta_1 + (\lambda\Delta x + 4)\zeta_2\right]\frac{\Delta t}{\Delta x} \tag{9.99}$$

$$\alpha_{4,4} = 2\left[3 + (u_1 + 2u_2)\frac{\Delta t}{\Delta x}\right] \tag{9.100}$$

$$c_1 = 3(2u_1 + u_2) + \left[2u_1^2 - u_1u_2 - u_2^2 + 3g(\zeta_1 - \zeta_2)\right]\frac{\Delta t}{\Delta x} - \frac{1}{2}(2u_1 + u_2)S_e \tag{9.101}$$

$$c_2 = \left\{\left[3h - \frac{1}{2}(h\lambda + S_o)\Delta x\right]u_1 - \left[3h + \frac{1}{4}(h\lambda + S_o)\Delta x\right]u_2\right\}\frac{\Delta t}{\Delta x}$$

$$+ \left\{\left(4 - \frac{3}{8}\lambda\Delta x\right)u_1\zeta_1 - \left(1 + \frac{1}{8}\lambda\Delta x\right)(u_1\zeta_2 + u_2\zeta_1) - \left(2 + \frac{1}{8}\lambda\Delta x\right)u_2\zeta_2\right\}\frac{\Delta t}{\Delta x}$$

$$+ 3(2\zeta_1 + \zeta_2) \tag{9.102}$$

$$c_3 = 3(u_1 + 2u_2) + \left[u_1^2 - u_1 u_2 - 2u_2^2 + 3g(\zeta_1 - \zeta_2)\right]\frac{\Delta t}{\Delta x} - \frac{1}{2}(u_1 + 2u_2)S_e \quad (9.103)$$

$$c_4 = \left\{\left[3h - \frac{1}{4}(h\lambda + S_o)\Delta x\right]u_1 - \left[3h + \frac{1}{2}(h\lambda + S_o)\Delta x\right]u_2\right\}\frac{\Delta t}{\Delta x}$$

$$+ \left\{\left(2 - \frac{1}{8}\lambda\Delta x\right)u_1\zeta_1 + \left(1 - \frac{1}{8}\lambda\Delta x\right)(u_1\zeta_2 + u_2\zeta_1) - \left(4 + \frac{3}{8}\lambda\Delta x\right)u_2\zeta_2\right\}\frac{\Delta t}{\Delta x}$$

$$+3(\zeta_1 + 2\zeta_2) \qquad\qquad (9.104)$$

where the energy loss term (S_e) is defined as

$$S_e = \frac{g|u_1 + u_2|\Delta t}{C_z^2\left(\dfrac{bh}{b + 2h}\right)} \qquad\qquad (9.105)$$

The resulting numerical scheme is implicit and requires solution of a system of equations by using a mathematical procedure, such as the conjugate gradient method (Beckman 1960).

By looking at the coefficients of matrix [A] and the constants in vector {C} obtained from the finite elements Galerkin method while applied to the telegrapher's equation (Equations 9.71 through 9.74 and Equations 9.75 and 9.76) and the Saint-Venant system (Equations 9.85 through 9.100 and Equations 9.101 through 9.104), it is evident that the finite elements method results into elaborately weighted finite differences schemes.

9.4 BOUNDARY ELEMENTS METHOD

The boundary elements method (BEM), also known as the boundary integral equation method (BIEM), is a very powerful numerical technique. However, the applicability of the method is limited to a specific class of partial differential equations. The main characteristic of the BEM is that the discretization into 'boundary elements' occurs only along the boundary, while the solution in the interior domain is obtained analytically (Partridge, Brebbia and Wrobel 1992).

9.4.1 Mathematical background

If a fluid flow is described by a vector field, then the divergence of the vector is a measure of the strength of a source or sink acting within the domain. Also, the conservation principle states that the integral of the domain's divergence must equal the flux of the vector field through the domains' boundary. Mathematically, this relation is known as the divergence theorem and it reads

$$\iiint_D \nabla \cdot \vec{H} \, d\omega = \iint_S \vec{H} \cdot \vec{n} \, ds \qquad (9.106)$$

where $\vec{H}$ is the vector field, $\vec{n}$ is the unit normal vector on the boundary and ω and s are the elementary volume and elementary boundary surface, respectively. Let us assume that the two functions G and F are twice differentiable in the domain D and that they satisfy the relations

$$\vec{H} = F\nabla G \qquad (9.107)$$

and

$$\vec{H} = G\nabla F \qquad (9.108)$$

Substitution of Equations 9.107 and 9.108 into Equation 9.106 results, respectively, in

$$\iiint_D [\nabla F \cdot \nabla G + F\nabla^2 G] \, d\omega = \iint_S F\nabla G \cdot \vec{n} \, ds \qquad (9.109)$$

$$\iiint_D [\nabla G \cdot \nabla F + G\nabla^2 F] \, d\omega = \iint_S G\nabla F \cdot \vec{n} \, ds \qquad (9.110)$$

By subtracting Equation 9.110 from Equation 9.109, the resulting equation is Green's second identity:

$$\iiint_D [F\nabla^2 G - G\nabla^2 F] \, d\omega = \iint_S (F\nabla G - G\nabla F) \cdot \vec{n} \, ds \qquad (9.111)$$

Furthermore, Equation 9.111 can be written as

$$\iiint_D [F\nabla^2 G - G\nabla^2 F]d\omega = \iint_S \left(F\frac{\partial G}{\partial n} - G\frac{\partial F}{\partial n} \right)ds \tag{9.112}$$

In case both functions G and F satisfy the Laplace equation ($\nabla^2 G = \nabla^2 F = 0$), then Equation 9.112 is reduced to the boundary integral

$$\iint_S \left(F\frac{\partial G}{\partial n} - G\frac{\partial F}{\partial n} \right)ds = 0 \tag{9.113}$$

and the solution is simplified substantially.

9.4.2 Boundary elements method in water resources

In water resources applications it is very common to deal with velocity potential fields. Thus, let F be the velocity potential φ, and G be a 'free space Green function' which satisfies the Laplace equation everywhere in the domain D, but at a singular point $P(x_i)$ function G goes to infinity. For a two-dimensional space, a free space Green function is given as

$$G = \ln(r) \tag{9.114}$$

where distance r is measured from point P. In order to apply Equation 9.113, the point P should be excluded by a small circle, σ, as shown in Figure 9.6. Then, after substituting the functions G and F, Equation 9.113 becomes a line integral (Power and Wrobel 1995):

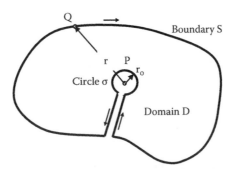

Figure 9.6 Exclusion of singular point P from the solution domain.

$$\int_S \left[\varphi \frac{\partial(\ln r)}{\partial n} - \ln r \frac{\partial \varphi}{\partial n} \right] ds + \lim_{r_0 \to 0} \int_\sigma \left[\varphi \frac{\partial(\ln r)}{\partial n} - \ln r \frac{\partial \varphi}{\partial n} \right] ds = 0 \qquad (9.115)$$

In addition, the integral around circle σ is estimated as

$$\lim_{r_0 \to 0} \int_0^{2\pi} \left[-\frac{\varphi}{r_0} + \ln r_0 \frac{\partial \varphi}{\partial r} \right] r_0 d\theta = -2\pi\varphi(P) \qquad (9.116)$$

Combining Equations 9.115 and 9.116 results in

$$2\pi\varphi(P) = \int_S \left[\varphi(Q) \frac{\partial(\ln r)}{\partial n} - \ln r \frac{\partial \varphi(Q)}{\partial n} \right] ds \qquad (9.117)$$

Therefore, the potential at any interior point P of the domain D can be estimated provided that both the potential φ and its derivative $\frac{\partial \varphi}{\partial n}$ on the boundary are known. In a well-posed problem, the values of either φ (Dirichlet condition), $\frac{\partial \varphi}{\partial n}$ (Von Neumann condition) or a combination of both (Robin condition) are provided. But in order to apply Equation 9.117, the challenge remains to estimate the values for the 'missing' boundary data. This can be accomplished by moving point P (base point) at the boundary and isolating it with a circular arc (Figure 9.7).

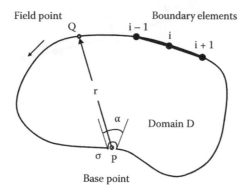

Figure 9.7 Base point and boundary elements.

Then Equation 9.117 becomes

$$\alpha\varphi(P) = \int_S \left[\frac{\varphi(Q)}{r} \frac{\partial r}{\partial n} - \ln r \frac{\partial \varphi(Q)}{\partial n} \right] ds \qquad (9.118)$$

where for a smooth boundary $\alpha = \pi$. The physical interpretation of Equation 9.118 is the balance between potential sources (lnr) and dipoles $\frac{\partial(\ln r)}{\partial n}$ weighted by $\frac{\partial \varphi}{\partial n}$ and φ, respectively.

Since either φ or $\frac{\partial \varphi}{\partial n}$ should be known, by selecting N-number of field points Q on the boundary, an N-number of equations is generated for the missing quantities. In summary the BEM procedure is as follows:

1. A finite number of field points Q_j (j = 1,N) are defined on the boundary.
2. A different point P_i (i = 1,N) is selected each time as the base point.
3. Linear boundary elements, along with local coordinate systems ξ-η, are defined between two successive points Q_j and Q_{j+1} (Figure 9.8).
4. The variable and its derivative are approximated by using appropriate shape functions.
5. Equation 9.118 is successively applied for all base points and all boundary elements.
6. The resulting algebraic system, depending solely on the geometry of the boundary, is then solved for the unknown values of φ and $\frac{\partial \varphi}{\partial n}$.

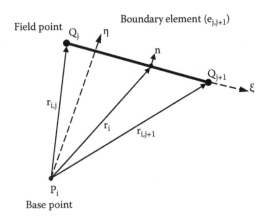

Field point Q_j η Boundary element ($e_{j,j+1}$)

n

Q_{j+1} ξ

$r_{i,j}$

r_i $r_{i,j+1}$

P_i
Base point

Figure 9.8 Boundary elements coordinate system ξ-η.

Typical linear approximations for the potential and its derivative in the region $\xi_j \leq \xi \leq \xi_{j+1}$ are as follows:

$$\hat{\varphi} = \frac{(\varphi_{j+1} - \varphi_j)\xi + (\xi_{j+1}\varphi_j - \xi_j\varphi_{j+1})}{\xi_{j+1} - \xi_j} \tag{9.119}$$

$$\frac{\partial \hat{\varphi}}{\partial n} = \left\{ \left[\left(\frac{\partial \varphi}{\partial n} \right)_{j+1} - \left(\frac{\partial \varphi}{\partial n} \right)_j \right] \xi + \left[\xi_{j+1} \left(\frac{\partial \varphi}{\partial n} \right)_j - \xi_j \left(\frac{\partial \varphi}{\partial n} \right)_{j+1} \right] \right\} \frac{1}{\xi_{j+1} - \xi_j} \tag{9.120}$$

Once both values of the potential and its derivative are known throughout the boundary, then Equation 9.117 can be used to calculate the potential within domain D. However, the interior solution is inaccurate near the boundary due to the fact that there is a discontinuity of the coefficient of φ from 2π in Equation 9.117 to α in Equation 9.118.

Concluding, it should be emphasized that the purpose of Chapter 9 – in addition to presenting a brief introduction of the weighted residual, the finite elements and the boundary elements methods – was to demonstrate in comparison the simplicity and efficiency of the finite differences method that was the main focus of this book.

References

Abbott, M.B. 1991. *Hydroinformatics: Information technology and the aquatic environment*. Aldershot, UK: Ashgate.

Abbott, M.B., and A.W. Minns. 1998. *Computational hydraulics*. Aldershot, UK: Ashgate.

Arvanitidou, S.K., K.L. Katsifarakis, and C.G. Koutitas. 2012. Comparative evaluation of two computational tools for flow simulation in zoned coastal aquifers. *Civil Engineering and Environmental Systems* 29(4): 273–281.

Bailard, J.A. 1981. An energetics total load sediment transport model for a plane sloping beach. *Journal of Geophysical Research* 96: 10938–10954.

Bear, J., A.H.-D. Cheng, S. Sorek, D. Ouazar, and I. Herrera, eds. 1999. *Seawater intrusion in coastal aquifers: Concepts, methods and practices*. Dordrecht, The Netherlands: Kluwer.

Bear, J., and A. Verruijt. 1987. *Modelling groundwater flow and pollution*. Dordrecht, The Netherlands: Reidel.

Beckman, F.S. 1960. The solution of linear equations by the conjugate gradient method. In *Mathematical methods for digital computers*, vol. 1, edited by A. Ralston and H.S. Wilf, 62–72. New York: Wiley.

Brebbia, C., and A.J. Ferrante. 1983. *Computational hydraulics*. London: Butterworths & Co.

Camemen, B., and M. Larsen. 2005. A general formula for non-cohesive bed load sediment transport. *Estuarine, Coastal and Shelf Science* 63: 249–260.

Chapman, S.J. 2013. *MATLAB programming with applications for engineers*. Australia: Cengage Learning.

Chau, K.W. 2010. *Modelling for coastal hydraulics and engineering*. New York: Spon Press.

Cheney, W., and D. Kincaid. 2013. *Numerical mathematics and computing*. 7th ed. Boston: Cengage Learning.

Chow, V.T. 2009. *Open-channel hydraulics*. Reprint. Caldwell, NJ: Blackburn Press.

Chung, T.J. 1978. *Finite element analysis in fluid dynamics*. New York: McGraw-Hill.

Copeland, G. 1958. A practical alternative to mild slope equations. *Coastal Engineering* 9(2): 125–149.

de Marsily, G. 1986. *Quantitative hydrogeology: Groundwater hydrology for engineers*. San Diego, CA: Academic Press.

Engelund, F., and E. Hansen. 1972. *A monograph on sediment transport in alluvial streams*. Copenhagen: Teknisk forlag. Technical University of Denmark.

Faires, J.D., and R.L. Burden. 2015. *Numerical methods*. 4th ed. Boston: Cengage Learning.

Fletcher, C.A.J. 1991. *Computational techniques for fluid dynamics: Fundamentals and general techniques*. Vol. 1, 2nd ed. Berlin: Springer-Verlag.

Fredshoe, J., and R. Deigaard. 1992. *Mechanics of coastal sediment transport*. Singapore: World Scientific.

Gelfand, I.M., and S.V. Fomin. 1963. *Calculus of variations*. Translated from Russian by R.A. Silverman. Englewood Cliffs, NJ: Prentice Hall.

Hromadka, T.V., B.L. Beech, and J.C. Clements. 1986. *Computational hydraulics for civil engineers*. Mission Viejo, CA: Lighthouse Publications.

Kamphuis, J.W. 1991. Incipient wave breaking. *Coastal Engineering* 15(3): 185–203.

Kim, Y.C., ed. 2009. *Handbook of coastal and ocean engineering*. Singapore: World Scientific.

Komen, G.J., L. Cavaleri, M. Donelan, K. Hasselmann, S. Hasselmann, and P.A.M. Janssen. 1994. *Dynamics and modelling of ocean waves*. Cambridge, UK: Cambridge University Press.

Kourafalou, V.H., Y.G. Savvidis, C.G. Koutitas, and Y.N. Krestenitis. 2004. Modelling studies on the processes that influence matter transfer on the Gulf of Thermaikos (NW Aegean Sea). *Continental Shelf Research* 24(2): 203–222.

Koutitas, C.G. 1983. *Elements of computational hydraulics*. London: Pentech Press.

Koutitas, C.G. 1988. *Mathematical models in coastal engineering*. London: Pentech Press.

Koutitas, C., and M. Gousidou-Koutita. 1986. A comparative study of three mathematical models for wind-generated circulation in coastal areas. *Coastal Engineering* 10(2): 127–138.

Koutitas, C., and M. Gousidou-Koutita. 2004. A model and a numerical solver for the flow generated by an air-bubble curtain in initially stagnant water. In *Proceedings of International Conference of Computational Methods in Science and Engineering (ICCMSE)*, 283–287. Athens, Greece.

Koutitas, C.G., M. Gousidou-Koutita, and V. Papazachos. 1983. A microcomputer code for tsunami generation and propagation. *Applied Ocean Research* 8(3): 156–163.

Lapidus, L., and G.F. Pinder. 1999. *Numerical solution of partial differential equations in science and engineering*. New York: Wiley-Interscience.

Lee, J., and C.S. Park. 2001. A weakly nonlinear wave model for practical use. In *Proceedings of 4th International Symposium on Ocean Wave Measurement and Analysis (ASCE)*, edited by B.L. Edge and J.M. Hemsley, 894–903, San Francisco, California.

Lencastre, A. 1995. *Hydraulique generale*. Paris, France: Eyrolles.

Lin, P. 2008. *Numerical modelling of water waves*. Abingdon, UK: Taylor & Francis.

Mitchell, A.R., and D.F. Griffiths. 1980. *The finite differences method in partial differential equations*. New York: John Wiley & Sons.

Moore, H., and S.K. Sanadhya. 2015. *MATLAB for engineers*. 4th ed. Boston: Pearson.

O'Connor, B.A., S. Pan, J. Nicholson, N. MacDonald, and D.A. Huntley. 1998. A 2D model of waves and undertow in the surf zone. In *Proceedings of 26th Conference on Coastal Engineering (ASCE)*, edited by B.L. Edge, 286–296. Copenhagen, Denmark.

Partridge, P.W., C.A. Brebbia, and L.C. Wrobel. 1992. *The dual reciprocity boundary element method*. Southampton, UK: Computational Mechanics Publications.

Pelnard-Considère, R., 1956. Essai de théorie de l'evolution des formes de rivages en plages de sable et de galets. *Quatrième Journées de l'Hydraulique, Les Energies de la Mer, Question III*, Rapport 1: 74-1-10.

Power, H., and L.C. Wrobel. 1995. *Boundary integral methods in fluid mechanics*. Southampton, UK: Computational Mechanics Publications.

Press, W.H., S.A. Teukolsky, W.T. Vetterling, and B.P. Flanney. 1992. *Numerical recipes: Example book (FORTRAN)*. 2nd ed. Cambridge, UK: Cambridge University Press.

Rezzolla, L. 2011. Numerical methods for the solution of partial differential equations. Lecture notes for the COMPSTAR School of Computational Astrophysics, 8-13/02/10, Caen, France. http://www.aei.mpg.de/~rezzolla/lnotes /Evolution_Pdes/evolution_pdes_lnotes.pdf.

Rodi, W. 2000. *Turbulence models and their applications in hydraulics: A state of the art review*. 3rd ed., 2nd repr. Rotterdam, The Netherlands: IAHR Monographs, Balkema.

Samaras, A.G., and C.G. Koutitas. 2008. Modelling the impact on coastal morphology of the water management in transboundary river basins: The case of River Nestos. *Management of Environmental Quality: An International Journal* 19(4): 455–466.

Savvidis, Y.G., M.G. Dodou, Y.N. Krestenitis, and C.G. Koutitas. 2004. Modelling of the upwelling hydrodynamics in the Aegean Sea. *Mediterranean Marine Science* 5(1): 5–18.

Scarlatos, P.D. 1982. A pure finite element method for the Saint-Venant equations. *Coastal Engineering* 6(1): 27–45.

Scarlatos, P.D. 1996a. Estuarine hydraulics. In *Environmental hydraulics*, edited by V.P. Singh and W.H. Hager, 289–348. Dordrecht, The Netherlands: Kluwer Academic.

Scarlatos, P.D. 1996b. Ecohydrodynamics. In *Environmental hydraulics*, edited by V.P. Singh and W.H. Hager, 349–397. Dordrecht, The Netherlands: Kluwer Academic.

Scarlatos, P.D. 2001. Computer modelling of fecal coliform contamination of an urban estuarine system. *Water Science and Technology, IWA* 44(7): 9–16.

Scarlatos, P.D. 2002. On the geometry of cohesive settling flocs. In *Fine sediment dynamics in the marine environment*, edited by J.C. Winterwerp and C. Kranenburg, 265–276. Amsterdam, The Netherlands: Elsevier.

Scarlatos, P.D., and A.J. Mehta. 1993. Instability and entrainment mechanisms at the stratified fluid mud-water interface. In *Nearshore and estuarine cohesive sediment transport*, edited by A.J. Mehta, 205–223. Washington, D.C.: American Geophysical Union.

Sharp, B.B., and D.B. Sharp. 1995. *Water hammer: Practical solutions*. Oxford, UK: Butterworth-Heinemann.

Shiflet, A.B., and G.W. Shiflet. 2014. *Introduction to computational science: Modeling and simulation for the sciences*. 2nd ed. Princeton, NJ: Princeton University Press.

Singh, V.P., and P.D. Scarlatos. 1988. Analysis of gradual earth-dam failure. *Journal of Hydraulic Engineering* 114(1): 21–42.

Sorensen, R.M. 2006. *Basic coastal engineering*. 3rd ed. New York: Springer Science.

U.S. Army Corps of Engineers. 2002. *Coastal engineering manual*. Engineer Manual 1110-2-1100, 6 vols. Washington, D.C.: U.S. Army Corps of Engineers.

Vreugdenhil, C.B. 1981. *Computational hydraulics: An introduction*. Heidelberg, Germany: Springer Verlag.

Yang, C.T. 2003. *Sediment transport: Theory and practice*. Reprint. Malabar, FL: Krieger.

Zafirakou-Koulouris, A., C. Koutitas, S. Sofianos, A. Mantziafou, M. Tzali, and S. Dermissi. 2012. Oil spill dispersion forecasting with the aid of a 3D simulation model. *Journal of Physical Science and Application* 2(10): 448–453.

Zienkiewicz, O.C. 1971. *The finite element method in engineering science*. New York: McGraw-Hill.

Index

Printed and bound by CPI Group (UK) Ltd, Croydon, CR0 4YY

01/11/2024

01782619-0011